GENE SILENCING: NEW RESEARCH

GENE SILENCING: NEW RESEARCH

GRACE W. REDBERRY
EDITOR

Nova Science Publishers, Inc.
New York

For permission to use material from this book please contact us:
Telephone 631-231-7269; Fax 631-231-8175
Web Site: http://www.novapublishers.com

Library of Congress Cataloging-in-Publication Data
Gene silencing : new research / Grace W. Redberry (editor).
p. ; cm.
Includes bibliographical references and index.
ISBN 1-59454-832-3
1. Gene silencing. I. Redberry, Grace W.
[DNLM: 1. Gene Silencing. QU 475 G326 2004]
QH450 .G4634
572.8'65--dc22 2005033773

Published by Nova Science Publishers, Inc. ✦ New York

Contents

Preface

Gene silencing is a general term describing epigenetic processes of gene regulation. The term gene silencing is generally used to describe the "switching off" of a gene by a mechanism other than genetic mutation. That is, a gene which would be expressed (turned on) under normal circumstances, is switched off by machinery in the cell. Genes are regulated at either the transcriptional or post-transcriptional level. Transcriptional gene silencing is the result of histone modifications, creating an environment of heterochromatin around a gene that makes it inaccessible to transcriptional machinery (RNA polymerase, transcription factors, etc.). Post-transcriptional gene silencing is the result of mRNA of a particular gene being destroyed. The destruction of the mRNA prevents translation to form an active gene product (in most cases, a protein). A common mechanism of post-transcriptional gene silencing is RNAi. Both transcriptional and post-transcriptional gene silencing are used to regulate endogenous genes. Interestingly, mechanisms of gene silencing also protect the organism's genome from transposons, viruses. This new book presents the latest research in this important field.

The genome-wide studies of gene expression with cDNA microarray make it easy to measure the transcription level of most genes at once. Various types of lymphomas/leukemias have been analyzed with the cDNA microarray to investigate the molecular basis of lymphomagenesis/leukemogenesis. In chapter I, the authors analyzed the expression pattern of the human NK/T cell line (NK-YS, NK-TY2) by cDNA-array and tissue-microarray comprehensively and systematically followed by RT-PCR and Western blotting. The significant changes in the gene expression of NK-YS and NK-TY2 cell line were detected: increase in 18 and decrease in 20 genes compared to those of normal NK cells or peripheral blood mononuclear cells. Among these genes, they found a strong decrease of hematopoietic cell specific protein-tyrosine-phosphatase *SHP1* mRNA. Further analysis with standard immunohistochemistry and tissue-microarray, which utilized 207 paraffin-embedded specimens of various kinds of malignant lymphomas, showed that 100% of NK/T lymphoma specimens and more than 95% of various types of malignant lymphoma were negative for SHP1 protein expression. The promoter region of the *SHP1* gene has been revealed to be highly methylated in patient samples of adult T cell leukemia/lymphoma (ATLL) (methylation frequency 90%), natural killer (NK)/T cell lymphoma (91%), diffuse large B-cell lymphoma (93%), MALT lymphoma (82%), mantle cell lymphoma (75%), plasmacytoma (100%) and follicular lymphoma (96%). The methylation frequency was significantly higher in high grade-MALT lymphoma cases (100%) than in low grade-MALT lymphoma cases (70%), correlating well with the frequency of no expression of SHP1 protein in high grade- (80%) and low grade-MALT lymphoma (54%). It suggests that the *SHP1* gene silencing with aberrant CpG methylation relates to the lymphoma progression in addition to the malignant transformation. Furthermore, the promoter methylation of the *SHP1* gene was clearly correlated with the clinical stage such as complete remission or relapse. Loss of heterozygosity with microsatellite markers near the *SHP1* gene was showed in 79% of informative ALL cases. These evidences show that the *SHP1* gene is a relevant novel biomarker of the wide range of hematopoietic malignancies. Additionally, our results suggest that the loss of the *SHP1* gene expression plays an important role in multistep lymphomagenesis/leukemogenesis.

Successful completion of the human genome project increased interest in mechanisms controlling gene expression. As described in chapter II, data presented by many research groups revealed silencing of genes in aging. Methylation in promoters regions, histone modification and RNAi mechanisms silence the genes which slow down the aging. Global DNA demethylation increases genomic instability, amplification of oncogenes and silencing of the genes through the RNAi mechanism, which accelerates the process. In aging, there is a decreased expression of tumor suppressors and genes involved in detoxification and cholesterol metabolism but, to the contrary, there is an increased expression of the oncogenes and the genes instrumental in chronic inflammation, stress response and silencing. Gene silencing affects crucial processes, such as cell cycle, metabolism, programmed cell death, nuclear transport, endogenous clocks and many others. Normal balance in gene expression, unlike mutations, can be restored by activation of silenced genes through the use of agents which decrease methylation of promoters, induce acetylation of histones and normalize global methylation of DNA. Small organic molecules, phenylacetylglutamine (PG) and phenylacetylisoglutamine (isoPG), increase expression of tumor suppressor genes *PTEN* and *MAD* and down regulate oncogenes *AKT2* and *MYCC*. Cosmetic cream of PG and isoPG was studied in clinical trial, which provided evidence of its antiaging efficacy on the skin. The reversal of gene silencing and down regulation of oncogenes are attractive targets in age management.

Epigenetic silencing of tumor suppressor genes plays an important role in the pathogenesis of human cancer. Especially, hypermethylation of the promoter regions of cancer-related genes is often observed in tumors. Recently, a novel tumor suppressor gene, which was termed Ras association domain family 1 (RASSF1A) gene was characterized by its frequent epigenetic silencing in human tumors as reported in chapter III. RASSF1 consists of two main variants (RASSF1A and RASSF1C), which are transcribed from distinct CpG island promoters. Aberrant methylation of the RASSF1A promoter region is one of the most frequent epigenetic inactivation events detected in human cancer and leads to silencing of RASSF1A. Silencing of RASSF1A was commonly observed in primary tumors including lung, breast, pancreas, kidney, liver, cervix, nasopharyngeal, prostate, thyroid and other cancers. Moreover, RASSF1A methylation was frequently detected in body fluids including blood, urine, nipple aspirates, sputum and bronchial alveolar lavages. Inactivation of RASSF1A was associated with an advanced tumor stage (e.g. bladder, brain, prostate, gastric tumors) and poor prognosis (e.g. lung, sarcoma and breast cancer). Detection of aberrant RASSF1A methylation may serve as a diagnostic and prognostic marker. The functional analyses of RASSF1A reveal an involvement in apoptotic signaling, microtubule stabilization and mitotic progression. The tumor suppressor RASSF1A may act as a negative Ras effector inhibiting cell growth and inducing cell death. Thus, RASSF1A represents a tumor suppressor gene, which is frequently epigenetically silenced in the pathogenesis of human cancers.

A prerequisite for understanding the functions and mechanisms of DNA methylation in mammals is knowledge of its distribution in the genome. However, the detailed long-range local methylation profiles are not available. The authors of chapter IV used bisulfite sequencing to obtain high-resolution methylation maps of the transcriptionally active mouse and human VHL loci. The VHL CpG island, including the promoter and the first exon, is not methylated in normal tissues, but hypermethylated in some sporadic renal carcinomas. Our maps cover 17 kb of human and 6.7 kb of mouse genomic DNA and include 434 and 151 CpG dinucleotides respectively. Despite the divergence of genomic sequences, similar methylation patterns were maintained in mouse and human cells. Both loci contained 1.5 kb methylation-protected regions, including a 1 kb GC-rich island and 0.5 kb of the adjacent part of intron 1. All other sequences, unique and repetitive alike, were heavily methylated. They then developed transgenic mice carrying the human VHL CpG island and the adjacent sequences to begin studying the nature of mouse and human DNA *cis* elements protecting these CpG islands from methylation. In 3 of 5 mouse strains originating from independent founder animals, methylation patterns of human transgenes resembled those of normal human cells indicating that the essential *cis* DNA protecting elements were present within the transgene and functional in the mouse environment. Developing mice, which form and maintain normal methylation patterns on human DNA templates, make possible systematic dissection of elements relevant for formation and maintenance of methylation patterns in tumor and normal cells.

Nearly 97% of the human genome is the non-coding DNA, which varies from one species to another, and changes in these sequences are frequently noticed to manifest clinical and circumstantial malfunction. Numerous non-protein-coding genes are recently found to encode microRNAs, which are responsible for RNA-mediated gene silencing through RNA interference (RNAi)-like pathways. MicroRNAs (miRNAs), small single-stranded regulatory RNAs capable of interfering with intracellular messenger RNAs (mRNAs) that contain either complete or partial complementarity, are useful for the design of new therapies against cancer polymorphism and viral mutation. This flexible characteristic is different from double-stranded siRNAs (small interfering RNAs) because a much more rigid complementarity is required for siRNA-induced RNAi gene silencing. MiRNAs were firstly discovered in *Caenorhabditis elegans* as native RNA fragments that modulate a wide range of genetic regulatory pathways during embryonic development. Currently, varieties of miRNAs are widely reported in plants, animals, and even microbes. Intronic microRNA is a new class of miRNAs derived from the processing of gene introns. The intronic miRNAs differ uniquely from previously described intergenic miRNAs in the requirement of type II RNA polymerases (Pol-II) and spliceosomal components for their biogenesis. Several kinds of Id-miRNAs have been identified in *C. elegans*, mouse and human cells; however, neither function nor application has been reported. In chapter V, the authors show for the first time that intron-derived miRNAs are able to induce RNA interference in not only human and mouse cells but also zebrafishes, chicken embryos and adult mice, demonstrating the evolutionary preservation of the intron-mediated gene silencing through miRNA functionality *in cell* and *in vivo*. These findings suggest an intracellular miRNA-mediated gene regulatory system, fine-tuning the degradation of protein-coding messenger RNAs.

Of all epigenetic modifications, hypermethylation is commonly convicted as a critical causal event in silencing important class of genes in cancer. In breast cancer, a series of tumor suppressor or related genes are frequently hypermethylated and this is associated with silencing of these genes. The authors now understand that DNA methylation machineries might include a number of DNA methyltransferases, proteins that direct DNA methyltransferases to specific promoters, chromatin modifying enzymes as well as demethylases. There is also data to suggest that pharmacological down regulation of some members of the DNA methylation machinery could inhibit cancer *in vitro, in vivo* and in clinical trials. The use of the methylation status of certain genes as a marker of cancer or cancerous conditions provides an additional weapon in early detection and prognosis of breast cancers. Chapter VI describes the techniques used to decipher the complement of methylation-inactivated genes, discuss the epigenetic regulation of certain critical tumor suppressor and growth regulatory genes implicated in breast cancer and its relevance to breast cancer development, progression and therapy.

In: Gene Silencing: New Research
Editor: Grace W. Redberry, pp. 1-32
ISBN 1-59454-832-3

Chapter I

High Frequent Gene Silencing of Hemetopoietic Cell Specific Protein Tyrosinephosphatase SHP1 in Hemetopoietic Cell Malignancies

Takashi Oka,[1,*] Mamoru Ouchida,[2] Mitsune Tanimoto,[3] Kenji Shimizu,[2] and Tadashi Yoshino[1]

[1]Department of Pathology, [2]Department of Molecular Genetics,
[3]Department of Hematology, Oncology and Respiratory Medicine, Okayama University Graduate School of Medicine and Dentistry, Okayama, Japan

Abstract

The genome-wide studies of gene expression with cDNA microarray make it easy to measure the transcription level of most genes at once. Various types of lymphomas/leukemias have been analyzed with the cDNA microarray to investigate the molecular basis of lymphomagenesis/leukemogenesis. Recently, our group analyzed the expression pattern of the human NK/T cell line (NK-YS, NK-TY2) by cDNA-array and tissue-microarray comprehensively and systematically followed by RT-PCR and Western blotting. The significant changes in the gene expression of NK-YS and NK-TY2 cell line were detected: increase in 18 and decrease in 20 genes compared to those of normal NK cells or peripheral blood mononuclear cells. Among these genes, we found a strong decrease of hematopoietic cell specific protein-tyrosine-phosphatase *SHP1* mRNA. Further analysis with standard immunohistochemistry and tissue-microarray, which utilized 207 paraffin-embedded specimens of various kinds of malignant lymphomas, showed that 100% of NK/T lymphoma specimens and more than 95% of various types of

* Correspondence: Takashi Oka Ph.D., DMSc., Department of Pathology, Okayama University Graduate School of Medicine and Dentistry, 2-5-1 Shikata-cho, Okayama 700-8558, Japan; Tel; +81- 86-235-7150; Fax; +81- 86-235-7156; e-mail; oka@md.okayama-u.ac.jp

malignant lymphoma were negative for SHP1 protein expression. The promoter region of the *SHP1* gene has been revealed to be highly methylated in patient samples of adult T cell leukemia/lymphoma (ATLL) (methylation frequency 90%), natural killer (NK)/T cell lymphoma (91%), diffuse large B-cell lymphoma (93%), MALT lymphoma (82%), mantle cell lymphoma (75%), plasmacytoma (100%) and follicular lymphoma (96%). The methylation frequency was significantly higher in high grade-MALT lymphoma cases (100%) than in low grade-MALT lymphoma cases (70%), correlating well with the frequency of no expression of SHP1 protein in high grade- (80%) and low grade-MALT lymphoma (54%). It suggests that the *SHP1* gene silencing with aberrant CpG methylation relates to the lymphoma progression in addition to the malignant transformation. Furthermore, the promoter methylation of the *SHP1* gene was clearly correlated with the clinical stage such as complete remission or relapse. Loss of heterozygosity with microsatellite markers near the *SHP1* gene was showed in 79% of informative ALL cases. These evidences show that the *SHP1* gene is a relevant novel biomarker of the wide range of hematopoietic malignancies. Additionally, our results suggest that the loss of the *SHP1* gene expression plays an important role in multistep lymphomagenesis/leukemogenesis.

Abbreviations

PTP	protein-tyrosine phosphatase;
PTPN6	protein-tyrosine phosphatase nonreceptor-type 6;
HCPH	hematopoietic cell phosphatase;
SHP	src homology region 2-domain phasphatase;
SH2	src-homology 2;
MSP	methylation specific polymerase chain reaction;
ALL	acute lymphoblastic leukemia;
AML	acute myeloid leukemia;
ATLL	adult T cell leukemia;
CML	chronic myelogenous leukemia;
DLBCL	diffuse large B-cell lymphoma;
MALT lymphoma	marginal zone lymphoma MALT type;
MCL	mantle cell lymphoma;
ALCL	anaplastic large cell lymphoma;
FL	follicular lymphoma;
RLH	reactive lymphoid hyperplasia;
PBMC	peripheral blood mononuclear cell;
NK	natural killer.

Introduction

As a consequence of the genetic instability or spontaneous mutation, malignant cells accumulate an increasing number of genetic aberrations during the course of tumor progression such as chromosomal rearrangements, deletions of anti-oncogenes, amplification or activation of oncogenes and various epigenetic changes that result in altered gene expression. Somatic mutations of the variable region of B- and T-lymphocyte antigen receptor genes have been shown to be a hallmark of lymphomas and leukemias, including germinal center B cells and their descendants. For several types of B-cell lymphomas, chromosomal translocations into immunoglobulin heavy chain switch regions have been described, such as translocation of c-*myc* in Burkitt's lymphomas (Magrath, 1990; Ferber et al., 2004), bcl-6 in diffuse large cell lymphomas (Ye et al., 1995; Ohno, 2004), bcl-2 in follicular lymphoma (Poetsh et al., 1996; Ladetto et al., 2003), bcl-1 in mantle–zone lymphoma (Komatsu et al., 1993; Brizard et al., 1997) and fibroblast growth factor receptor-3 in multiple myeloma (Chesi et al., 1997; Martinelli et al., 2001). Such kinds of conserved genetic changes for these own lineage have been revealed, however universally conserved genetic changes and gene expression alterations allover malignant lymphoma/leukemia have not been identified yet.

Completion of the genome sequences of model organisms and humans will provide us with complete blueprints of these genomes (Lander et al., 2001; Venter et al., 2001). The study of gene expression on a genomic scale is the most obvious opportunity made possible by complete genome sequences, and the most experimentally straightforward. cDNA microarrays make it easy to measure the transcripts for every gene at once (Zhang et al., 1997; Duggan et al., 1999; Wang K. et al., 1999; Alizadeh et al., 2000; Nakanishi et al., 2001). Due to the tight connection between the function of a gene product and its expression regulation, each gene is expressed in specific cells and under specific conditions. Promoter of a gene functions as a transducer, responding to input information about the identities, environment and internal state of a cell by changing the level of transcription of the gene. The group of genes expressed in a cell determines what the cell is made of, what biochemical and regulatory systems are operative, how the cell is built and what it can and cannot do. Many genes and signal transduction pathways that control cellular proliferation, differentiation and programmed cell death, as well as genomic integrity, are also involved in cancer development. cDNA microarrays have enabled measurement of the expression of thousands of genes in a single experiment for the systematic and comprehensive exploration of genome-wide alterations in cancer cells (Zhang et al., 1997; Wang K. et al., 1999; Alizadeh et al., 2000). Recently, distinct types of diffuse large B-cell lymphoma (DLBCL), which show markedly different clinical courses and treatment responses apparently reflecting the variation in tumor proliferation rate, host response and differentiation state of tumor, have been identified and diagnosed by gene expression profiling with cDNA microarrays (Alizadeh et al., 2000). However, analysis of hundreds of specimens from patients from different stages is essential to establish the diagnostic, prognostic and therapeutic importance of each of the many oncogene or anti-oncogene candidates. To overcome this difficulty, an array-based high-throughput tissue microarray technique has been established to facilitate analysis of gene expression and DNA copy number of large numbers of tumors (Kononen et al., 1998; Schraml et al., 1999; Moch et al., 1999). We developed this technique independently, modified, and here apply the modified method to lymphoma analysis in combination with a cDNA expression array (Oka et al., 2001).

In the recent investigation, we analyzed the expression profiles of human NK cell lines compared to normal NK cells or peripheral blood mononuclear cells (PBMCs) to elucidate the mechanism of NK/T lymphomagenesis with cDNA and tissue microarrays. We found strong suppression or abolishment of the *SHP1* gene expression in the almost of all hematopoietic malignancies as well as in NK/T cell lymphomas by both genetic and epigenetic mechanisms (Oka et al., 2001; Oka et al., 2002; Koyama et al., 2003). Herein, we review the recent progress of the pathological role and the therapeutic potential of the *SHP1* gene in lymphomas/leukemias and related diseases.

SHP1 Protein

The growth and many functional responses of hematopoietic cells are regulated through the phosphorylation of tyrosine residues and the dephosphorylation of phosphotyrosine residues in proteins. The gene encoding the cytoplasmic protein-tyrosine phosphatase SHP1 (also termed PTP1C, SH-PTP1, HCP and PTPN6) is commonly expressed in hematopoietic cells. SHP1 is a 68-kDa protein of a non-receptor type protein-tyrosine phosphatase containing two tandem *Src* homology 2 (SH2) domains, a catalytic domain, and a C-terminal tail of about 80 amino acid residues (Shen et al., 1991) (Figure 1). The SH2 domains act as phosphotyrosine binding domains, and mediate the interaction of this protein-tyrosine phosphatase (PTP) with its substrates. The catalytic domain catalyzes the dephosphorylation of phosphotyrosine peptides. The other protein-tyrosine phosphatase, which is highly homologous with SHP1, is SHP2 (also termed PTP2C, PTP1D, SH-PTP2, SH-PTP3, Syp and PTPN11) (Figure 2). SHP2 is widely expressed in most tissues and plays a regulatory role in various cellular signaling events containing mitogenic activation, metabolic control, transcriptional regulation and cell migration. SHP2 mutations have been detected in the patients with Noonan syndrome, LEOPARD syndrome and acute myeloid leukemia (Tartaglia et al., 2001; Legius et al., 2002; Digilio et al., 2004; Loh et al., 2004; Yoshida et al., 2004). X-ray crystal structure analysis of SHP1 exhibited the association of the important amino acids on each domain and the function in detail. Yang et al. showed that the residues in the WPD loop, especially the amino acid following Asp421, are critical for the movement of WPD loop on binding substrates and the specific activity of protein-tyrosine phosphatase (Yang J. et al., 1998), and showed that the catalytic domain of SHP1 prefers the substrate with the consensus sequence (D/E)X(L/I/V)XnpYXX(L/I/V) (n=1 or 2) (Yang et al., 2000). It has been reported that ligand-free SHP1 has an auto-inhibited conformation, and that the N-terminal SH2 (N-SH2) domain blocks the catalytic domain and keeps the enzyme in the inactive conformation, which supports that the phosphatase activity is primarily regulated by the N-SH2 domain (Pei et al, 1994; Yang et al., 2003). Binding of phosphopeptide to the C-SH2 domain results in large conformational changes that restore the distorted conformation of the neighboring N-SH2 domain and subsequently opens up its phosphopeptide-binding pocket to harbor a second phosphopeptide molecule (Yang et al., 2003) (Figure 3). It has been thought that the C-terminal tail can also modulate the phosphatase activity, because an SHP1 truncation mutant which lacked the last 35 amino acids of the C-terminal tail showed a 30-fold increase in activity relative to wild-type SHP1 (Pei et al, 1994).

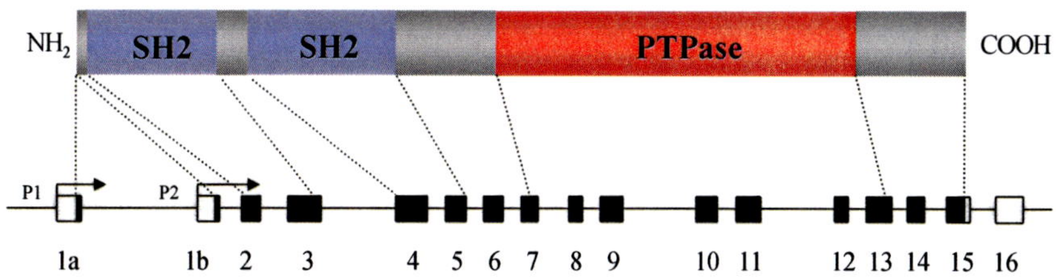

Figure 1. Schematic representation of the *SHP1* gene and SHP1 protein. SHP1 protein has two SH2 domains and catalytic (PTPase) domain. White boxes represent the non-coding exons, and black boxes represent the coding exons. Exon number is shown below boxes. *P1*, promoter for exon 1a expressed in non-hematopoietic cells; *P2*, promoter for exon 1b expressed in hematopoietic cells.

```
SHP1  MLSRGWFHRDLSGLDAETLLKGRGVHGSFLARPSRKNQGDFSLSVRVGDQVTHIRIQNSG   60
SHP2  MTSRRWFHPNITGVEAENLLLTRGVDGSFLARPSKSNPGDFTLSVRRNGAVTHIKIQNTG   60
                         SH2 domain
SHP1  DFYDLYGGEKFATLTELVEYYTQQQGVLQDRDGTIIHLKYPLNCSDPTSERWYHGHMSGG  120
SHP2  DYYDLYGGEKFATLAELVQYYMEHHGQLKEKNGDVIELKYPLNCADPTSERWFHGHLSGK  120

SHP1  QAETLLQAKGEPWTFLVRESLSQPGDFVLSVLSDQPKAGP-GSPLRVTHIKVMCEGGRYT  179
SHP2  EAEKLLTEKGKHGSFLVRESQSHPGDFVLSVRTGDDKGESNDGKSKVTHVMIRCQELKYD  180
                         SH2 domain
SHP1  VGGLETFDSLTDLVEHFKKTGIEEASGAFVYLRQPYYATRVNAADIENRVLELNKKQESE  239
SHP2  VGGGERFDSLTDLVEHYKKNPMVETLGTVLQLKQPLNTTRINAAEIESRVRELSKLAETT  240

SHP1  DTAKAGFWEEFESLQKQEVKNLHQRLEGQRPENKGKNRYKNILPFDHSRVILQGRDSNIP  299
SHP2  DKVKQGFWEEFETLQQQECKLLYSRKEGQRQENKNKNRYKNILPFDHTRVVLHDGDPNEP  300

SHP1  GSDYINANYIKNQLLGP---DENAKTYIASQGCLEATVNDFWQMAWQENSRVIVMTTREV  356
SHP2  VSDYINANIIMPEFETKCNNSKPKKSYIATQGCLQNTVNDFWRMVFQENSRVIVMTTKEV  360
                         Catalytic domain
SHP1  EKGRNKCVPYWPEVGMQRAYGPYSVTNCGEHDTTEYKLRTLQVSPLDNGDLIREIWHYQY  416
SHP2  ERGKSKCVKYWPDEYALKEYGVMRVRNVKESAAHDYTLRELKLSKVGQGNTERTVWQYHF  420

SHP1  LSWPDHGVPSEPGGVLSFLDQINQRQESLPHAGPIIVHCSAGIGRTGTIIVIDMLMENIS  476
SHP2  RTWPDHGVPSDPGGVLDFLEEVHHKQESIMDAGPVVVHCSAGIGRTGTFIVIDILIDIIR  480

SHP1  TKGLDCDIDIQKTIQMVRAQRSGMVQTEAQYKFIYVAIAQFIETTKKKLEVLQSQKGQES  536
SHP2  EKGVDCDIDVPKTIQMVRSQRSGMVQTEAQYRFIYMAVQHYIETLQRRIEEEQKSKRKGH  540

SHP1  EYGNITYPPAMKNAHAKASRTSSKHKEDVYENLHTKNKREEKVKKQRSADKEKSKGSLKRK 597
SHP2  EYTNIKYSLADQTSGDQSPLPPCTPTPPCAEMREDSARVYENVGLMQQQKSFR         593
```

Figure 2. Alignment of amino acid sequence of SHP1 and SHP2 proteins. Identical amino acids are indicated by red and similar residues are indicated by pink. Two SH2 domains and catalytic domain are shown in boxes.

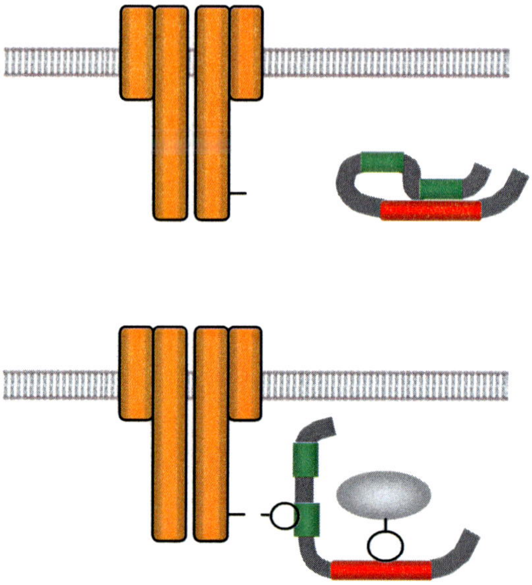

Figure 3. Model of the activation of SHP1. The N-SH2 domain blocks the catalytic domain and keeps the enzyme in the inactive conformation. Binding of phosphopeptide to the C-SH2 domain results in large conformational change and subsequently opens up its phosphopeptide binding pocket to harbor a second phosphopeptide molecule (Yang et al., 2003). *PTPase*, catalytic domain.

Genomic Structure

The *SHP1* gene consists of 17 exons spanning about 17kb on 12p13 of human chromosome (Plutzky et al., 1992; Matsushita et al., 1999). This gene has two alternative promoters (Figure 1 and 4). Exon 1a is expressed in non-hematopoietic cells and exon 1b is exclusively expressed in cells of the hematopoietic lineage. The exon 1a and 1b encode Met Leu Ser Arg Gly and Met Val Arg from translation initiation codon, respectively. The orthologous genes are identified as the *Hcph* gene on mouse chromosome 6 and the *Ptph6* gene on rat chromosome 4p42. Genomic structure of the *SHP1* gene is almost similar in human, mouse and rat, and the nucleotide length of each coding exon is also conserved over species (Figure 4). Interestingly, the promoter region of the exon 1a is commonly located at about 500b downstream of the gene rich cluster *C10* gene in human, mouse and rat genome DNA (Figure 5). The expression of the *C10* gene may affect to the expression of the exon 1a of the *SHP1* gene. The hematopoietic specific-promoter region for exon 1b is shown in Figure 6. The proximal region is highly conserved between human, mouse and rat. Nucleotide motif analysis suggests that binding sequences for some transcriptional factors are located on the promoter region (Figure 7) as well as previously reported (Banville et al., 1995; Tsui et al., 2002). Some alternatively spliced variants of the *SHP1* gene have been reported. Banville et al. showed minor transcripts from which part of or the entire N-SH2 domain coding sequences are missing in nonhematopoietic cells (Banville et al., 1995). Jin et al. showed a C-terminal alternatively spliced form of SHP1, which is 29 amino acids longer than SHP1, and the 66 C-terminal amino acids of which are different from SHP1 (Jin et al., 1999). Beghini et al. found a variant derived from aberrant splicing within the N-SH2 domain leading to retention of intron 3, in acute myeloid leukemia patients. The level of the aberrant intron-retaining splice variant was lower in CD117 +/- AML bone marrow mononuclear cells at remission than at diagnosis, suggesting an involvement of posttranscriptional processing of the SHP1 transcripts in leukemogenesis (Beghini et al., 2000). Ma et al. reported a deletion transcript found only in Jurkat leukemia-derived cells, which is predicted to encode for a 7-kDa protein containing most of the amino-terminal SH2 domain, and found an 88-nucleotide insert that is the unspliced second intron resulting in a frame-shift and the formation of a noncoding transcript (Ma et al., 2003). These findings support the concept that SHP1 can negatively regulate growth of malignant cells.

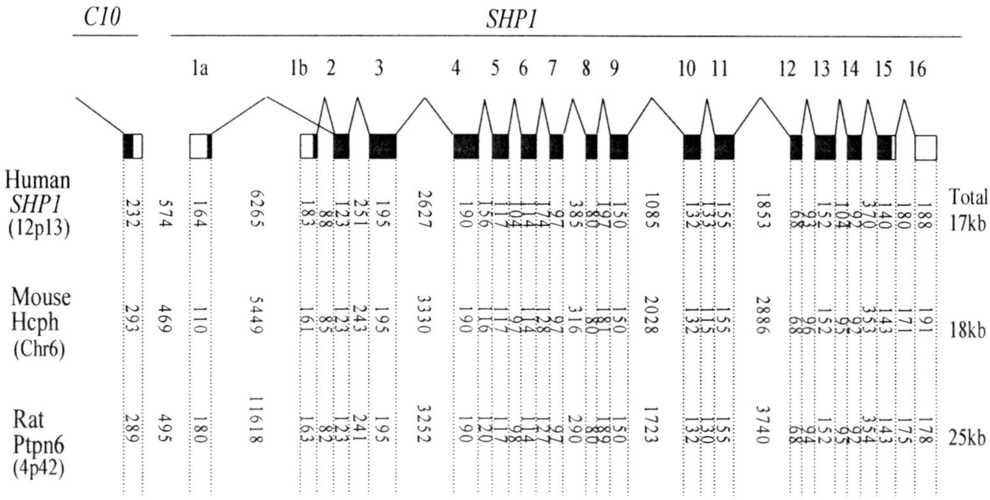

Figure 4. Genomic structure of the *SHP1* gene in human, mouse and rat. White boxes indicate non-coding exons, and black boxes indicate coding exons. Exon number is shown above the exons. Nucleotide numbers of exon, intron and the flunking region are shown under the schematic representation of exon-intron. Exon 3 of the gene rich cluster C10 gene is shown at 5' upstream of the *SHP1* gene.

Roles of SHP1 in the Hematopoietic Cells

SHP1 was found to be an important negative regulator of antigen receptor-mediated signaling and selection in B-lymphocytes (Cyster and Goodnow, 1995; Pani et al., 1995). SHP1 also has a significant role in the regulation of T-cell responses to antigen receptor-mediated stimulation, with its effects on signaling pathways evoked by T-cell receptor (TCR)

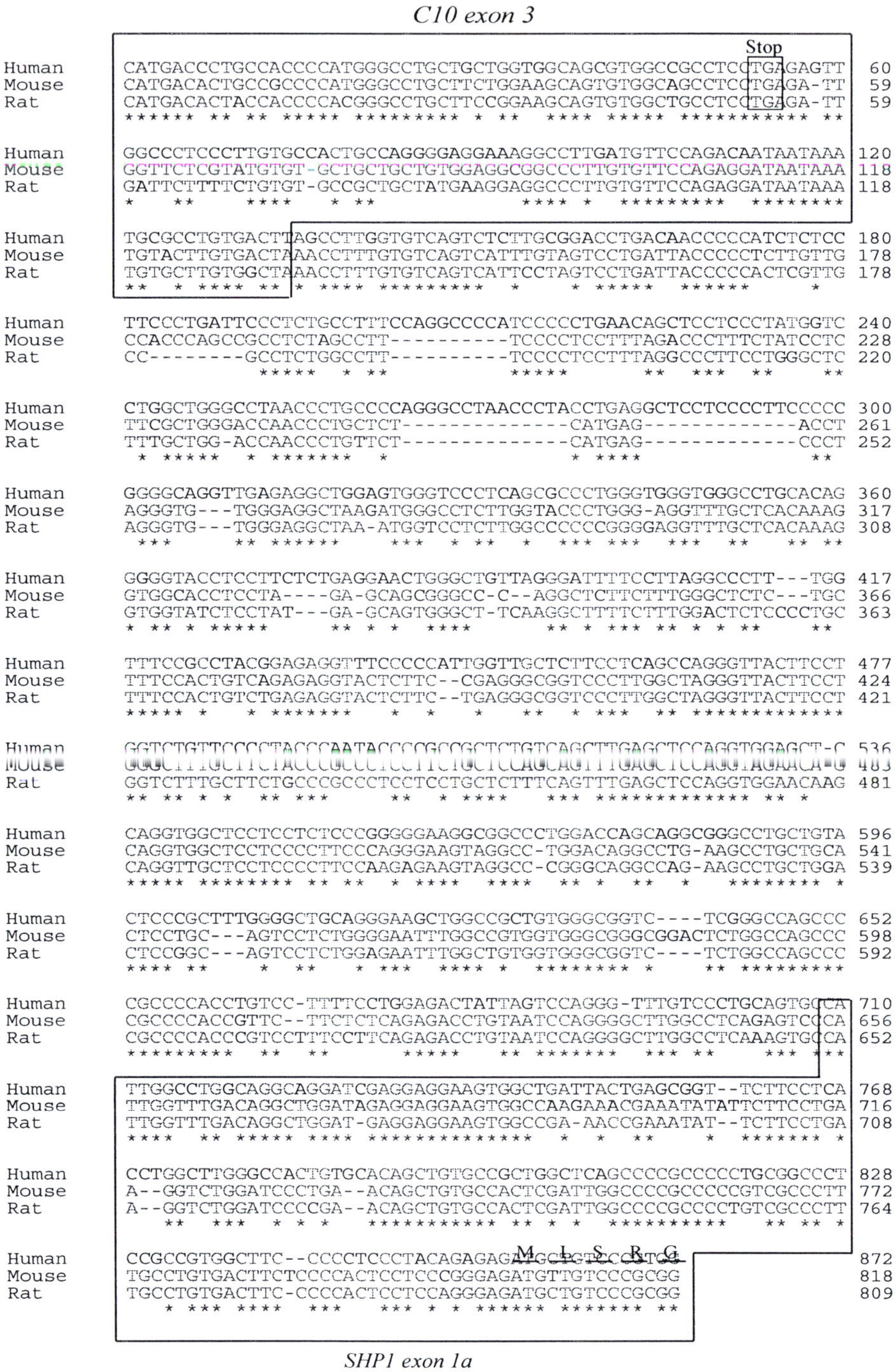

Figure 5. Alignment of the nucleotide sequences of promoter for the *SHP1* exon 1a in human, mouse and rat. Shown are the regions from 3' half of the *C10* gene exon 3 to the *SHP1* gene exon 1a. Identical nucleotides between two or three species are indicated by red. Asterisks indicate the completely identical nucleotides over three species. First five amino acids, which are encoded in the exon 1a of the human *SHP1* gene, are shown above the codons.

```
Human    ATCTCTGTTCTTTGAAGCCCTGAGGGCTGGTGTATAGGAGTTCAAAG-CACTGGCTTTGG  59
Mouse    ACTACTGTTCTTTGG-GTCCTGAGGGCTGGCAAGCAGAATTCAAAAGTCATTGGCTTTGA  59
Rat      ACTACTGTTCTTTGG-GACCTGAGGGTTGGCAAGCAGAATTCAAAAATCATTGGTTTTGA  59
         *   **********  * ******** ***     ** * *  ***  ** *** ****

Human    AACCGGACTGTCTGGGTTTGAATCCTGGCACTGCAGCTGACTCACTGATGGACTCAGG-- 117
Mouse    AAC-----TATTTAGGTCTGAATCCTG------CAACTCACTAAGGGTAAGACCTGGGGG 108
Rat      AAC-----TATTTAGGTCTGAGTCCTG------CCACTCACTAAGGGTAAGGCCTGGGG- 107
         ***     * * * *** *** *****      *  ** *** *  *   * *   **

Human    CAATGCCTTAAACTCCCTGAGCCTCAGGTTCCTTGTCTGTAAAATG---ATAAAGATAGC 174
Mouse    CAATTCCTCAAACTTCTT-GGCCTCAGTTTCCTTGTCTGTGAAGTGGGATAAAGATAGT 167
Rat      CAGTTCCTCAGACTTCTC-AGCCTCAGTTTCCTTGTCTATAAAGTGGGTATAAAGATAGT 166
         ** * *** * *** *    ******* ********** * ** **   **********

Human    CCCTGTTTCATAGGGCTGTGGTGAGAAACCAATCAGACAAGGCATGTGAACGCCATTATA 234
Mouse    GCCTGAGGAATGAGGCCCTGGTGACCTGTAACTCACAGGAGGCATGAAAATGACATTGTA 227
Rat      ACCTGTGTAGTGAGGCCGTGGTGCCCTGTTACTCACAGGTGGCATGTAATTGACATTTGA 226
          ****     *  ***  *****       * *** *   ******  *  * ****  *

Human    GCACAGCGCCCGGCATCCAGCAGGACTC-ACTCGATGACAGTTGTCACCGC-CATCATTG 292
Mouse    GCAATGAGTCTGGCAGGCTGCGACACTCCATTAAATGACGGT-GCCACTGCACACGATCA 286
Rat      CCAGAGAGTCTGGCATGCGTTGACACTCCATTAAATGACAGT-GTCACCACACACTGTCA 285
          **  * * * ****  *      **** * *  ***** ** * ***  * **   *

Human    TTATTA--GCGTGGGCCAGGGAGG---GCTGCGTAAAAGCAGCTGGTGGAGGAGGGAGAG 347
Mouse    TTATTA--GACTGTGCCTAAGAATAATTCTGAGTAGAGGCAGCTGCTGGAGGAGGAAGGG 344
Rat      GTATTATTAGCTGTGCCTTAGAAT---TCTGAGTAGAAGCAGCTGCTGGAGGAGGAAAGG 342
          *****     ** ***   **      *** *** * ******* ********* *  *

Human    ATGCCGTGGGACCGTCTGGGTTCGC-ATGCGTGAAGTATTATCTGGGCCTGGAGTGTGCA 406
Mouse    TTGACCTGATAATGTGGATATTCACTGTGCGTGACTCTTGACCTGC-------------- 390
Rat      TTGACCTGATCATGTGGATATTCACTGTGTGTGACTCTTGACCTGGAG--GGACTGAGGG 400
          ** * **     **     *** *  ** ****    * * ***

Human    AGGCACACATGTGTCCTTACTGCATGTGTTGTCACATATGTGCAATGCCATGCTCCTGAG 466
Mouse    ---CACACGTGTG-CCTTACCTCTTGAGTCATGA---ATATTTACTGCCTTGCCCCTGTG 443
Rat      GCACCCACGTGTGTCCTTACCTCTCGAGTCATGA---GTGTTTAGTGCCATGCTCCTGTG 457
            * *** **** ******  *  * **  * *    * *  * **** *** **** *

Human    CCTTTGATTGCAGACGTGTGGGAAGTGGGCCCCGTCCCCACCCCCAGTGCCACCCTGCTC 526
Mouse    CCTTTGTTGGCAGAC-TGTGGGAAGTGGGCCCT-CCCCCACCC----TGCCACCCT--CC 495
Rat      CCTCTGATGGCAGAC-TGTGCGAAGTGGGCCCT-CCCCCACCC----TGCCACCCT--TC 509
         *** ** * ****** **** ***********   ********    *********   *

Human    TGCTTCTCTTCCCTTGCTGTGCTCTAAAACGAGAAGTACAAGTGAGTTCCCCCAAGGGGT 586
Mouse    AGCTCCTCTTCCCTCGCT-----CTGAAAAGAGAAGTTCTAGTGAGTTCCCCTAAGGGGT 550
Rat      AGCTTCTCTTCCCTTGCT-----CTGAAAAGAGAAGTACAAGTGAGTTCCCCTAAGGGGT 564
          *** ********* ***     ** *** ******* * ************ *******

Human    CGGCCGCGCCTCTTCCTGTCCCCGCCCTGCCGGCTGCCCCAGGCCAGTGCAGTGGCAGCC 646
Mouse    CGGCCGCGCCCCTTCCTGTCTCCGCCCTGCTGGCTGCCCCAGGCCAGTAGAGTGGTAGCC 610
Rat      CGGCCGCGCCCCTTCCTGTCTCCGCCCTGCTGGCTGCCCCAGGCCAGTAGAGTGGTAGCC 624
         ********** ********* ********* *****************  ***** ****

Human    CCAGAACTGGGACCACCGGGGGTGGTGAGGCGGCCCGGCACTGGGAGCTGCATCTG--AG 704
Mouse    CGGGGAGCAGGACTGCAGGTTGGCTTTGGAGGCCTGGGCTCTGAGAGCCTTGCCTG--AG 668
Rat      CGGG-AGCAGGACAGCAGGTTGGCTTTGGAGGCCTGGGCTCTGAGAGCCTTGAGTAGAAG 683
         *  * *   ****  * **  *   *  *  * *  *** *** ****      *   **

Human    GCTTAGTCCCTGAGCTCTC---TGCCTGCCCAGACTAGCTGCACC-TCCTCATCCCCTGC 760
Mouse    GCTCATCTCTAGAGTTTGTACGTGCCTGCCCAGACAAACTGTTCCCTCCACATTTTCTGC 728
Rat      GCTCAGCTCTAGAGTTTTTAGGTGCCTGCCCAGACAGACTGTTCCCTCCACGTTTTGTGC 743
         *** *   *  *** *      *************   ***  ** *** * *    ***
                                              M  V  R
Human    GCCCCCTTCCTCTCCGGAAGCCCCCAGGATGGTGAG                         796
Mouse    AGCCAATTCAGT----GAGAACCCCAGGATGGAGAG                         760
Rat      AGCCAATACACC----AAGAGCCTCAGGATGGTGAG                         775
           **  * *        *   ** ************
```

SHP1 exon 1b

Figure 6. Alignment of the nucleotide sequences of promoter for the *SHP1* exon 1b in human, mouse and rat. The *SHP1* gene exon 1b and the proximal upstream region are shown. Identical nucleotides between two or three species are indicated by red. Asterisks indicate the completely identical nucleotides over three species. First three amino acids, which are encoded in the exon 1b of the human *SHP1* gene, are shown above the codons.

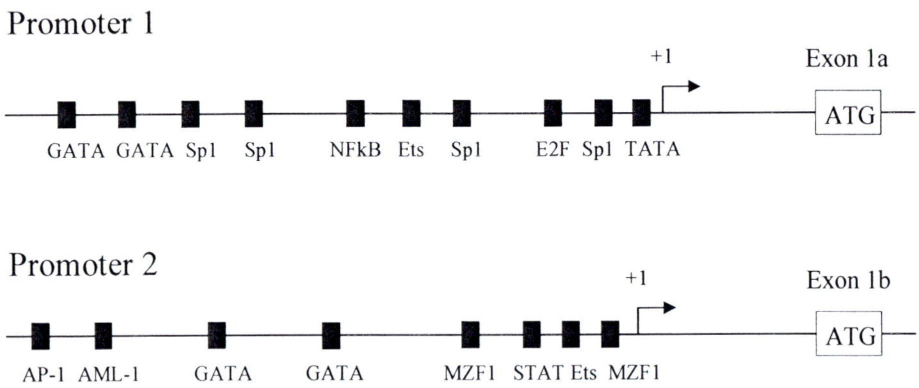

Figure 7. Schematic representation of the two promoter regions of the human *SHP1* gene. Transcriptional initiation sites are indicated by +1. Translational initiation codons are indicated by ATG. Black boxes indicate the putative binding sequence for transcriptional factors. GATA, GATA box; Sp1, Sp1 binding site; NFkB, NFκB binding site; Ets, ETS binding site; E2F, E2F binding site; TATA, TATA box, AP-1, AP-1binding site; AML-1, AML-1 binding site; MZF1, myeloid zinc finger protein MZF1 binding site; STAT, STAT binding site.

engagement predominantly inhibitory (Pani et al., 1996; Lorenz et al., 1996a; Carter et al., 1999). SHP1 is known to be associated with multiple signaling molecules, including ZAP70 (Healy and Goodnow, 1998), CD3ε (Brockdorff, 1999; Pani et al., 1996), CD5 (Pani et al., 1996) and interleukin-2R (Migone et al., 1998) in T cells; interleukin-3 receptor β chain (Yang W. et al., 1998) and erythropoietin receptor (Bittorf et al., 1999) in hematopoietic cells; CD22 (Blasioli et al., 1999), B-cell receptor (Cornall et al., 1999), SLP76 (Binstadt et al., 1998; Mizuno et al., 1996) and CD72 (Wu et al., 1998; Fusaki et al., 2000) in B cells; and KIR (killer cell inhibitory receptor) (Wang LL. et al., 1999) in natural killer cells. These interactions appear to exert primarily inhibitory effects on their signaling cascades. SHP1 is thought o be capable of terminating an activating signal by dephosphorylating some molecules early involved in signal transduction. Mice with the recessive motheaten (me) or the allelic viable motheaten (me^{v}) mutations, carrying mutaions in the *Hcph* gene (the *SHP1* mouse orthologous gene) express a severe autoimmune and immunodeficiency syndrome. Tsui et al. showed that the basic defect involves lesions in the gene that encodes hematopoietic cell phosphatase (Tsui et al., 1993). Shultz et al. showed that 2 allelic motheaten mutations result in aberrant splicing of the *Hcph* transcript (Shultz et al., 1993). The homozygous (me/me) or (me^{v}/me^{v}) mice exhibited multiple hematopoietic abnormalities including neutrophilia, lymphopenia, splenomegaly and/or elevated serum immunoglobulin, severe combined immunodeficiency and systemic autoimmune diseases, due to the disregulation of hematopoietic cell development (Tsui et al., 1993; Shultz et al., 1993; Lorenz et al., 1996b). Thus, motheaten was the first animal model for a specific protein-tyrosine phosphatase deficiency, useful in determining the precise role of HCPH in hematopoiesis. However little is known about the functional role of SHP1 in the malignant transformation process in the hematopoietic cells.

cDNA Microarray and Tissue Microarray Analyses of Lymphoma/Leukemia

In order to investigate the lymphomagenesis and leukemogenesis, we comprehensively and systematically analyzed the mRNA and protein expression pattern in the human lymphoma cell lines by cDNA-array and tissue-microarray, respectively (Oka et al., 2001). Initially we focused on NK/T-cell lymphoma, because frequent cases of nasal/nasopharyngeal NK/T-cell lymphoma have been reported in Asian and South American countries; these were characterized as polymorphous pleomorphic morphologies with azurophilic granules, the immunophenotypic profile of $CD2^+$, $CD3^-$, $CD56^+$, lack of rearranged T-cell receptor genes, and association with the Epstein-Barr virus (EBV)(Jaffe et al., 1996). These tumors are regarded as one of the most aggressive lymphoma types currently known because of their strong resistance against third generation combination chemotherapy and rapid, unfavorable clinical course. In order to investigate the molecular pathological mechanism of NK/T cell lymphoma, we analyzed the expression profiles of human NK cell lines (NK-YS and NK-TY2) which retained original characters of NK/T lymphoma. We detected some genes which showed significant changes in the gene expression of these cell lines. Among these genes, we found a strong decrease of *SHP1* mRNA by cDNA-array and RT-PCR (Figure 8) (Oka et al., 2001).

The Aberrant Expression of SHP1 Protein in Malignant Lymphoma and Leukemia Cell Lines

We next investigated the expression of SHP1 protein in various types of hematopoietic cell lines (Oka et al., 2001) and summarized with the results of other investigators (Table 1), showing that SHP1 protein expression is greatly suppressed or abolished in highly aggressive lymphomas including NK/T lymphoma, adult T-cell leukemia/lymphoma (ATLL) and Burkitt's lymphomas.

In the case of T-cell lymphoma, all human T-lymphotropic virus type-I (HTLV-I) positive tumor cell lines established from ATLL patient specimen (EDS, ATL1K, C91PL, C10MJ2, MT1, ATL2 and ILTMot) and four out of five HTLV-I (-) cell lines originated from cutaneous T-cell lymphoma (CTCL) (HUT78, SEZ4, 2A and 2B) showed completely negative for SHP1 protein expression. Human cord blood T cell lines (MT2 and MT4) or human peripheral blood T-lymphocytes lines (IWA3, MB3-12A and MB3-12B), which were established by the infection of HTLV-I with the co-culture of virus producer cells or ATLL patient mononuclear cells, showed abolished or decreased expression of SHP1 protein except for IWA1 cells. IWA1 is a sister cell line of IWA3 cells, which are originated from the peripheral blood lymphocytes of the identical donor and established at the same time by the co-cultivation with the HTLV-I high producer cell line, MT2 cells. Both cell lines are HTLV-I producer cells with normal human karyotype. Only the difference of them is the IL2–dependency; IWA1 is IL2-dependent, on the other hand, IWA3 is IL2 independent. The constitutive activation of some side-pathway in IWA1 cells is suggested to overcome the down-regulating control of the strong expression of SHP1, which is an interesting issue to be investigated. Like NK-YS cells, another NK/T cell lymphoma cell line NK-TY2 and anaplastic large cell lymphoma cell line (ALCL)(SUDHL1) also showed no expression of SHP1 protein.

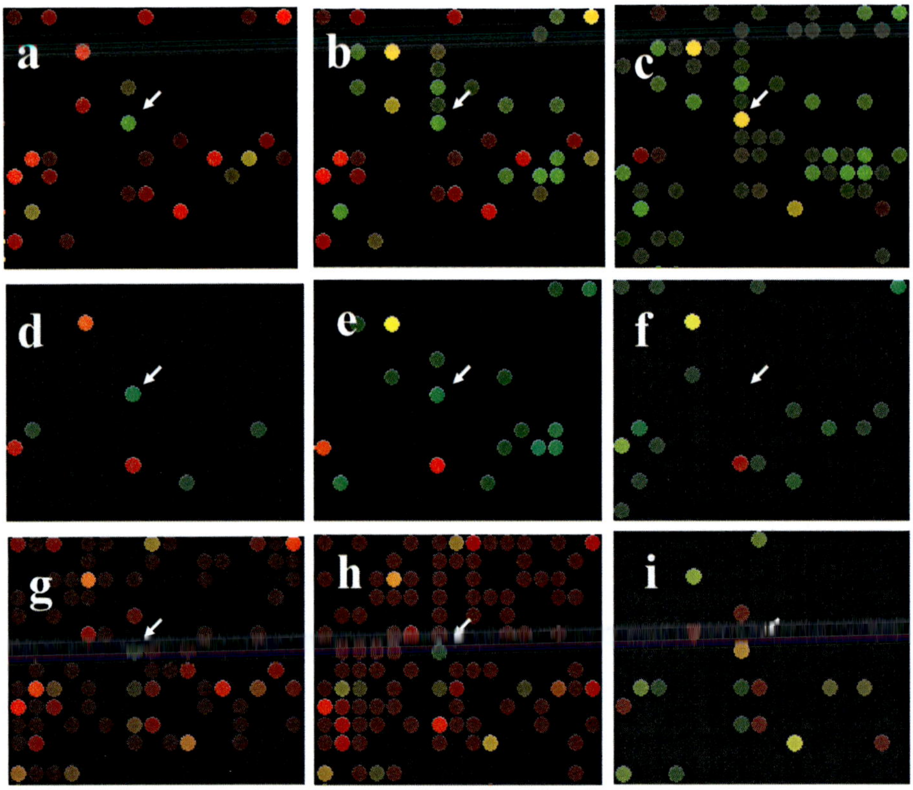

Figure 8. cDNA expression array analysis of NK-YS, NK-TY2, Jurkat, freshly isolated human PBMCs and human NK cells. Expression profile of NK-YS cells compared to that of fresh normal human NK cells (a), normal human PBMCs (b) and Jurkat cells (g). cDNA expression array profile of NK-TY2 were shown as the control of fresh normal human NK cells (d), PBMCs (e), NK-YS cells (f) and Jurkat cells (h). Expression profile of fresh NK cells compared to that of normal human PBMCs (c). Expression profile of Jurkat cells was shown as the control of PBMCs (i). Arrows indicate the green spots in (a), (b) and (g), showing that *SHP1* gene expression of NK-YS cells decreased compared with fresh NK cells (a), PBMCs (b) and Jurkat cells (g). Arrows indicate green spots in (d), (e) and (h) showing the reduced expression of *SHP1* gene in NK-TY2 cells compared with fresh NK cells (d), PBMCs (e) and Jurkat cells (h). Yellow spots indicated by arrows in (c) and (i), showing that *SHP1* gene expression of fresh NK cells (c) and Jurkat cells (i) were comparable to that of normal human PBMCs. Black spot in (f) pointed by arrow, showing that no expression of the *SHP1* gene in NK-TY2 cells is comparable to no expression of SHP1 gene in NK-YS cells. (Oka et al., 2001)

Table 1. Expresssion of SHP1 protein in the cell lines originated from hematopoietic malignancies

cell line	origin	SHP1	cell type	Ref.	cell line	origin	SHP1	cell type	Ref.
EDS	ATLL	-	T cell HTLV-I(+)	1,3	HUT78	SS	-	T cell	5
ATL1K	ATLL	-	T cell HTLV-I(+)	6	SEZ4	CTCL	-	T cell	2
C91PL	ATLL	-	T cell HTLV-I(+)	7	PB-1	CTCL	+	T cell	2
C10MJ2	ATLL	-	T cell HTLV-I(+)	7	2A	CTCL	-	T cell	2
MT1	ATLL	-	T cell HTLV-I(+)	2,3,4	2B	CTCL	-	T cell	2
ATL2	ATLL	-	T cell HTLV-I(+)	7	Jurkat	T-ALL	+	T cell	6
ILTMot	ATLL	-	T cell HTLV-I(+)	3	SKW3	T-CLL	-	T cell	6
MT2	cord blood	-	T cell HTLV-I(+)	4	SUDHL1	ALCL	-	null cell	2
MT4	cord blood	-	T cell HTLV-I(+)	4	NK-YS	NK/T-LY	-	NK/T	3,6,8
IWA1	PBL	+++	T cell HTLV-I(+)	6,8	NK-TY2	NK/T-LY	-	NK/T	3,6,8
IWA3	PBL	±	T cell HTLV-I(+)	3,8					
MB3-12A	PBL	-	T cell HTLV-I(+)	4					
MB3-12B	PBL	-	T cell HTLV-I(+)	4					
MJ	CTCL	+	T cell HTLV-I(+)	4					
HUT102B2	CTCL	+	T cell HTLV-I(+)	7,4					
Ramos	Burkitt's LY	-	B cell EBV(-)	1,6,8	RaeI	Burkitt's LY I/II	↓	B cell EBV(+)	1
WMN	Burkitt's LY	↓	B cell EBV(-)	1	Eli	Burkitt's LY I/II	↓	B cell EBV(+)	1
BL70	Burkitt's LY	-	B cell EBV(-)	1	Daudi	Burkitt's LY I/II	-	B cell EBV(+)	1,6,8
BL2	Burkitt's LY	↓	B cell EBV(-)	1	Raji	Burkitt's LY I/II	↓	B cell EBV(+)	1
BL28	Burkitt's LY	↓	B cell EBV(-)	1	Motu I	Burkitt's LY I/II	↓	B cell EBV(+)	1
BL41	Burkitt's LY	↓	B cell EBV(-)	1	Kem I	Burkitt's LY I/II	↓	B cell EBV(+)	1
BL47	Burkitt's LY	↓	B cell EBV(-)	1	OKUBL	Burkitt's LY I/II	↓	B cell EBV(+)	1
CA46	Burkitt's LY	↓	B cell EBV(-)	1	BL74	Burkitt's LY I/II	↓	B cell EBV(+)	1
EW6	Burkitt's LY	↓	B cell EBV(-)	1	ODHIBL	Burkitt's LY I/II	↓	B cell EBV(+)	1
DG75	Burkitt's LY	↓	B cell EBV(-)	1	BL72	Burkitt's LY I/II	↓	B cell EBV(+)	1
MC116	Burkitt's LY	↓	B cell EBV(-)	1	WW2L72	Burkitt's LY I/II	-	B cell EBV(+)	1

Table 1. Expresssion of SHP1 protein in the cell lines originated from hematopoietic malignancies (Continued)

cell line	origin	SHP1	cell type	Ref.	cell line	origin	SHP1	cell type	Ref.
Louckes	Burkitt's LY	↓	B cell EBV(-)	1	AG876BL	Burkitt's LY III	+	B cell EBV(+)	1
BJAB	Burkitt's LY	+	B cell EBV(-)	1	KemIII	Burkitt's LY III	+	B cell EBV(+)	1
DHBL	Burkitt's LY	↓	B cell EBV(-)	1	K124	Burkitt's LY III	+	B cell EBV(+)	1
BL72/LCL	normal LCL	+	B cell EBV(+)	1	W1BL	Burkitt's LY III	+	B cell EBV(+)	1
BL74/LCL	normal LCL	+	B cell EBV(+)	1					
DHBL/LCL	normal LCL	+	B cell EBV(+)	1					
SU-DHL4	B-LY(DLBCL)	+	B cell EBV(-)	6	HDLM	HD	-		2
BALL1	B-ALL	+	B cell	6,8	HDLM2	HD	-		6,8
SCOTT	B-ALL	-	B cell	6,8	L428	HD	-		6,8
KW	B-ALL	-	B cell	6,8	HD70	HD	+		6
KCA	B-ALL	+	B cell	6,8	L540	HD	-		2
SP53	MCL	+	B cell EBV(-)	6,8	MOLP2	MM	++		6,8
SP50B	MCL	+	B cell EBV(+)	6,8	RPMI8226	MM	-		6,8
SP49	MCL	+	B cell	8	U266	MM	-		6,8
HairM	HCL	+	B cell	6					
K562	CML	-		3,6,8	KG1	AML	+		6,8

ALL, acute lyphoblastic leukemia; ATLL, adult T cell leukemia/lymphoma; AML, acute myeloblastic leukemia; ALCL, anaplastic large-cell lymphoma;CLL, chronic lymphocytic leukemia; CML, chronic myelogenous leukemia; HCL, hairy cell leukemia; HD, Hodgkin's disease; LCL,lymphoblastoid cell line; LY, lymphoma; MM, multiple myeloma;CTCL, cutaneous T-cell lymphoma; SS, Sezary syndrome; Burkitt's LY I,II,III,Burkitt's lymphoma latency I,II,III; PBL;periphral blood lymphocyte; MCL,mantle cell lymphoma; NK/T LY, natural killer/T lymphoma; DLBCL,diffuse large B-cell lymphoma; Ref 1,Delibrias et al., (1997); Ref 2, Zhang et al.,(2000); Ref 3,Oka et al.(2002); Ref 4,Migone et al.,(1998), Ref 5,Leon et al.,(2002); Ref 6,Oka et al.,(2001); Ref 7, Zhang et al.(1999); Ref 8, Koyama et al.,(2003)

In the case of B-cell lymphoma, the SHP1 protein expression in most of EBV-positive and EBV-negative Burkitt's lymphoma (one of the high grade B-cell lymphoma) cell lines was decreased or abolished (Delibrias et al., 1997; Oka et al., 2001). Four out of five Hodgkin Disease (HD) cell lines (HDLM, HDLM2, L428 and L540) and two out of three multiple myeloma (MM) cell lines (RPMI18226 and U266) showed no expression of SHP1 protein. However, EBV-positive latency III Burkitt's lymphoma cell lines (AG876BL, KemIII, KK124 and WW1BL) expressed SHP1 protein as do the normal peripheral T or B cells and also normal lymphoblastoid cell lines (LCL) established with EBV infection (Delibrias et al., 1997), suggesting SHP1 expression is closely related EBV-latency.

For the leukemia cell lines, one T-cell chronic lymphocytic leukemia (T-CLL) cell line (SKW3), and two out of four B-cell acute lymphocytic leukemia cell lines (SCOTT and KW) and one chronic myelogenous leukemia (CML) cell line (K562) showed no expression of SHP1 protein. On the other hand, one T-cell acute lymphocytic leukemia (T-ALL) cell line (Jurkat), one hairy cell leukemia (HCL) cell line (HairM), mantle cell lymphoma cell lines (SP53 and SP50B), diffuse large B-cell lymphoma (DLBCL) cell line (SU-DHL4) expressed SHP1 protein normally (Oka et al., 2001). It is not clear why the *SHP1* gene is expressed in the different manner in these cell lines. At least, highly aggressive tumor cell lines such as ATLL, NK/T lymphomas and Burkitt's lymphomas are consistently abolished the *SHP1* gene expression, which is well coincide with the results of patient specimen. As discussed in the following section, epigenetic events such as DNA methylation have been revealed one of the fundamental mechanism of the *SHP1* gene silencing, one possible explanation of the different results of these cell lines is the reversed expression of SHP1 protein with the some events causing de-methylation of the CpG island of the *SHP1* gene in the artificial environmental condition of the tissue culture, which should be examined by the further investigation.

The Aberrant Expression of SHP1 Protein in Malignant Lymphoma and Leukemia Patient Specimens

In order to clarify whether the decreased mRNA expression of the *SHP1* gene in NK/T cell lymphoma cell lines *in vitro* is also observed in NK/T cell lymphoma patient specimens and to determine whether this phenomenon is specific to NK/T cell lymphoma or common among various types of lymphomas, we analyzed with standard immunohistochemistry and tissue-microarray, which utilized 207 paraffin-embedded specimens of various kinds of malignant lymphomas, showed that SHP1 protein was decreased in most of the malignant lymphomas; 100% of NK/T cell lymphoma and more than 95% of malignant lymphoma/leukemia patient specimens of DLBCL, follicular lymphoma (FL), Hodgkin lymphoma (HL) (Hodgkin's disease (HD)), mantle cell lymphoma (MCL), peripheral T-cell lymphoma (PTCL), ATLL and plasmacytoma were negative in SHP1 protein expression. Kossev et al. reported the consistent results that most of follicular lymphoma (29/30; 97%) were negative for SHP1 protein expression (Kossev et al., 2001). On the other hand, SHP1 protein was strongly expressed in mantle and marginal zones and some of interfollicualr zone lymphocytes in normal reactive lymphoid hyperplasia (RLH) specimens and resting peripheral blood lymphocytes which were used as positive control (Figure 9, Table 2). These immunohistochemical data was confirmed also by the *SHP1* mRNA in situ hybridization, showing the same results as the immunohistochemistry (Figure 9. e, f).

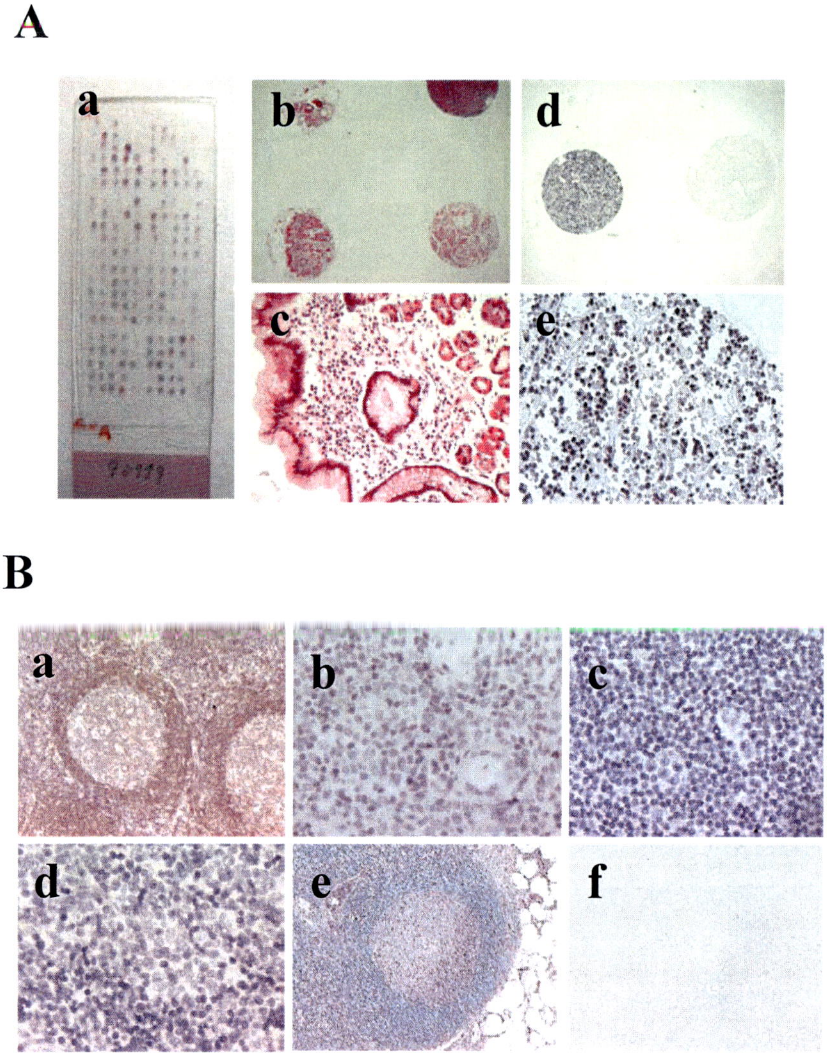

Figure 9. Tissue microarray analysis of SHP1 immunostaining and the *SHP1* mRNA in situ hybridization. (A) Hematoxylin-eosin (HE) staining and EBER1 mRNA *in situ* hybridization. An overview of a tissue microarray section containing 207 tissue samples (a); HE staining of tissue microarray in higher magnification (b, c); EBER1 mRNA *in situ* hybridization showing positive (left) and negative (right) reaction (d); higher magnification of EBER1-positive sample (e). (B) Immunohistochemistry of SHP1 protein, showing reactive lymphocyte hyperplasia (RLH) (a); NK/T cell lymphoma (b); Hodgkin's disease (c) and mantle cell lymphoma (d). *SHP1* mRNA in situ hybridization of reactive lymphocyte hyperplasia (e) and NK/T lymphoma (f). (Oka et al., 2001)

Table 2. Expression of SHP1 protein and CpG island methylation of the *SHP1* gene in malignant lymphoma/leukemia patient specimens

Diagnosis	Subtype	SHP1 protein*(%)		-	*SHP1* gene**	Refe-
		++	□		methylation(%)	rence
ATLL		0/9***(0)	0/9(0)	9/9(100)	18/20(90.0)	1,6
NK/T		0/11(0)	0/11(0)	11/11(100)	10/11(90.9)	1,6
DLBCL	Nodal	1/12(8.3)	0/12(0)	11/12(91.7)	23/26(88.5)	1
	Extranodal	0/12(0)	0/12(0)	12/12(100)	19/19(100)	1
	Total	1/24(4.2)	0/24(0)	23/24(95.8)	42/45(93.3)	1
MZL		3/13 (23.1)	2/13 (15.4)	8/13 (61.5)	n.d.	2
MALT lymphoma	Low grade	3/26(11.5)	9/26(34.6)	14/26(53.9)	7/10(70)	1
	High grade†	0/10(0)	2/10(20)	8/10(80)	7/7(100)	1
	Total	3/36(8.3)	11/36(30.6)	22/36(61.1)	14/17(82.4)	1
MCL		1/13(7.7)	0/13(0)	12/13(92.3)	3/4(75)	1
Plasmacytoma		0/17(0)	0/17(0)	17/17(100)	5/5(100)	1
FL	Grade 1	0/6(0)	0/6(0)	6/6(100)	11/11(100)	1
	Grade 2	0/6(0)	0/6(0)	6/6(100)	9/10(90)	1
	Grade 3	0/2(0)	0/2(0)	2/2(100)	3/3(100)	1
	Total	0/14(0)	0/14(0)	14/14(100)	23/24(95.8)	1
HD		0/22 (0)	0/22 (0)	22/22 (100)	n.d.	2
PT		1/20 (5)	0/20 (0)	19/20 (95)	n.d.	2
ALL		n.d.	n.d.	n.d.	5/8(62.5)	6
AML		n.d.	n.d.	n.d.	9/10(90)	6
CML		n.d.	n.d.	n.d.	11/11(100)	6
PBMCs	Healthy donor	11/11(100)	0/11(0)	0/11(0)	0/11(0)	1
RLH	Germinal center	0/10(0)	0/10(0)	10/10(100)	n.d.	1
	Mantle zone	10/10(100)	0/10(0)	0/10(0)	n.d.	1
	Interfollicular	6/10(60)	4/10(40)	0/10(0)	n.d.	1
MM		n.d.	n.d.	n.d.	27/34(79%)	3
ALCL		n.d.	n.d.	n.d.	36/44(82%)	4
MCL		n.d.	n.d.	n.d.	11/13(85%)	5
FL		n.d.	n.d.	n.d.	9/9(100%)	5

* detected by immunohistochemical staining with tissue microarray and standard immunohistochemistry.
** detected by methylation-specific PCR.
*** the number of expression-positive or methylation-positive samples / the number of total samples.
† DLBCL with MALT lymphoma, DLBCL, diffuse large B-cell lymphoma; MALT lymphoma, marginal zone lymphoma MALT type; MCL, mantle cell lymphoma; ALCL, anaplastic large cell lymphoma, FL, follicular lymphoma; PBMCs, peripheral blood mononuclear cells; RLH, reactive lymphoid hyperplasia; n.d., not done, Ref 1;Koyama et al., 2003, Ref 2;Oka et al., 2001, Ref 3; Chim et al., 2004a, Ref 4;Khoury et al., 2004, Ref 5;Chim et al., 2004b, Ref 6; Oka et al., 2002.

In comparison, about 60% of low grade malignant lymphoma such as marginal zone lymphoma (MZL) and MALT-lymphoma were negative for SHP1 protein expression (Oka et al., 2001). Kossev et al. reported that some malignant B-cell lymphomas such as MCL, MZL and chronic lymphocytic leukemia/small lymphocytic lymphoma were reported to express SHP1 protein at the normal level (Kossev et al., 2001). These low grade lymphomas are possibly in the intermediate state to become the decreased expression of SHP1 protein with either by the CpG island methylation of the *SHP1* promoter region or by the post-transcriptional blockage, causing the somewhat different results with the minor difference of the staining condition or pretreatment of the specimen.

The *SHP1* Gene Silencing with Aberrant Methylation of CpG Island

Cancer can be regarded as an epigenetic disease as well as a genetic disease. The genomes of transformed cells undergo simultaneous global genomic hypomethylation and dense hypermethylation of CpG islands associated with gene regulatory regions (Brown and Strathdee, 2002; Das and Singal, 2004). The reciprocal relationship between the density of methylated cytosine residues and intensity of a gene expression has been widely documented (Razin and Cedar, 1991; Brown and Strathdee, 2002; Das and Singal, 2004). Dramatic changes in methylation are thought to lead to chromosomal instability, loss of imprinting, activation of endogenous viruses, illegitimate expression, aneuploidy, mutation, and may also contribute to the epigenetic transcriptional repression of tumor suppressor genes in a wide variety of tumor types (Brown and Strathdee, 2002; Esteller and Herman, 2002).

To investigate a mechanism for the reduced expression of the *SHP1* gene, we analyzed the promoter region for the *SHP1* exon 1b which is known to express with hematopoietic specific manner. By methylation specific PCR (MSP) and unmethylation specific PCR, CpG island of the *SHP1* promoter was revealed to be highly methylated in large number of lymphoma/leukemia cell lines such as NK-YS, NK-TY2, K562, Daudi, HDLM2, L428, U266, RPMI18226, ATL1K and EDS cell lines and weakly methylated in SCOTT, KW, SP50B, and Ramos cell lines. (Figure 10 and Table 3). Chim et al. also reported the hypermethylation of the *SHP1* gene in HL60, U937, Raji and AF10 cells (Chim et al., 2004b). Additionally, hypermethylation of the CpG island of the *SHP1* gene was detected high frequently in various types of lymphoma/leukemia patient samples, including diffuse ATLL (methylation frequency 90%), NK/T lymphoma (91%), large B-cell lymphoma (DLBCL) (93%), MALT lymphoma (82%), mantle cell lymphoma (75%), plasmacytoma (100%) and follicular lymphoma (96%) by methylation-specific PCR (Fig.11 and Table 2). Patient specimens from MM (methylation frequency; 79%) and ALCL (82%) were also revealed to be highly methylated in the CpG island of the *SHP1* gene (Table 2) (Chim et al., 2004a; Khoury et al., 2004). In contrast, no MSP signals were detected in normal PBMCs. Unmethylation-specific PCR revealed that normal PBMCs and some cell lines, which express SHP1 mRNA and protein, showed strong unmethylation-specific signals. The ratio of methylation/unmethylation signals in the promoter CpG island correlate well with the reduced *SHP1* gene expression (Table 2). The methylation frequency was significantly higher in high-grade MALT lymphoma cases (100%) than in low-grade MALT lymphoma cases (70%), which was correlate well with the frequency of no expression of SHP1 protein in high-grade (80%) and low-grade MALT lymphoma (54%), suggesting that the *SHP1* gene silencing with aberrant CpG methylation relates to the lymphoma progression. In order to address the question whether the lymphoma cells themselves have totally methylated *SHP1* DNA and contaminated with the inflammatory cells or infiltrating normal PBMCs supplying

Table 3. SHP1 mRNA/protein expression and CpG island methylation of the promoter region of the *SHP1* gene in hematopoietic cel lines

Cell line	Origin	SHP1 gene methylation			SHP1 gene	SHP1 protein	SHP1 mRNA	Refe-
		RE-PCR	MSP	unMSP	rearrangement*	expression**	expression***	rence
NK-YS	NK/T-Ly	+++	+++	–	–	–	–	1,2
NK-TY2	NK/T-Ly	+++	+++	–	–	–	–	2
K562	CML	++	+++	–	–	–	–	1
IWA1	T(HTLV-1)	–	–	+	–	+++	++	1
KCA	B-ALL	–	–	+	–	+	+	1
BALL1	B-ALL	–	–	+	–	+	+	1
SCOTT	B-ALL	–	+	+	–	–	–	1
KW	B-ALL	–	+	++	–	–	–	1
SP53	MCL	–	–	++	–	+	++	1
SP50B	MCL	+	±	++	–	+	++	1
SP49	MCL	–	–	++	–	+	++	1
Ramos	Burkitt Ly	+	+	–	–	–	–	1
Daudi	Burkitt Ly	++	++	–	–	–	–	1
HDLM2	HD	+	++	–	–	–	–	1
L428	HD	+	+++	–	–	–	–	1
U266	MM	+++	+++	–	–	–	–	1
MOLP2	MM	–	–	++	–	++	++	1
RPMI8226	MM	++	+++	+	–	+	+	1
PBMCs	Healthy volunteer	–	–	+	–	+	+	1
ATL1K	ATLL	nd	++	++	–	–	+	2
EDS	ATLL	nd	+	++	–	–	+	2
Jurkat	T-ALL	nd	+	+	–	+	+	2,3
HL60	PML	nd	–	++	nd	nd	nd	3
U937	DHL	nd	++	++	nd	nd	nd	3
Raji	Burkitt's Ly	nd	++	++	nd	nd	nd	3
AF10	-	nd	+	++	nd	nd	nd	3

The strength of the SHP1 bands was indicated according to the intensity from +++, ++, +, to ±. – indicates completely negative for SHP1 band.

*:detected by Southern blot analysis **:detected by Western blot analysis ***:detected by RT-PCR PBMCs, peripheral blood mononuclear cells; Ly, lymphoma; CML, chronic myelogenous leukemia; T(HTLV-1), HTLV-1-immortalized normal human T cell; ALL, acute lymphoblastic leukemia; MCL, mantle cell lymphoma; HD, Hodgkin's disease; MM, multiple myeloma; DHL,diffuse histiocytic lymphoma PML, promyelocytic leukemia; RE-PCR, restriction enzymes-mediated polymerase chain reaction; MSP, methylation-specific polymerase chain reaction; unMSP, unmethylation-specific polymerase chain reaction

Reference 1, Koyama et al, 2003; Ref 2, Oka et al.2002; Ref 3, Chim et al., 2004b

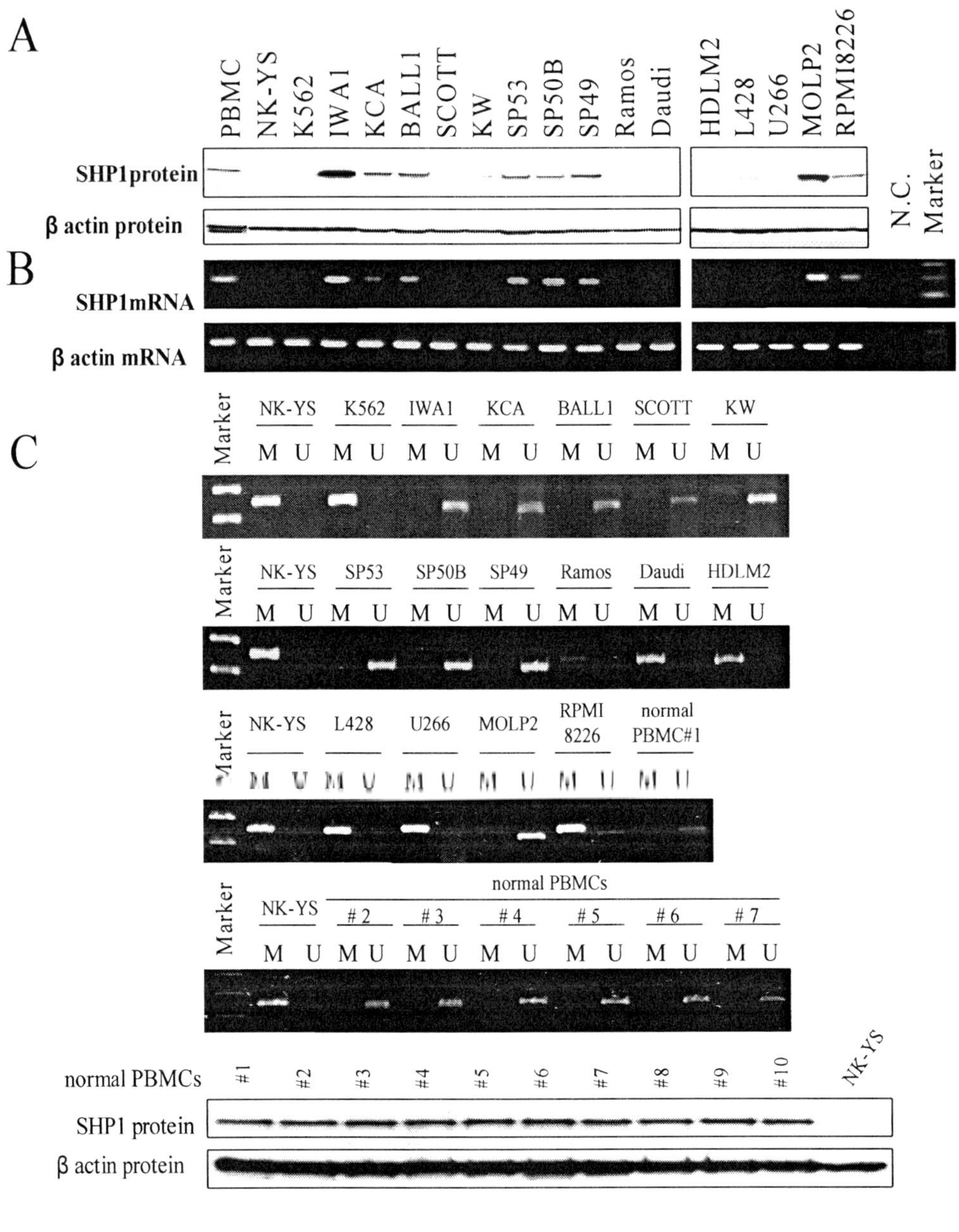

Figure 10. Analysis of the *SHP1* gene expression and methylation in various hematopoietic cell lines. (A) Western blot analysis of SHP1 protein expression. Cell lysates extracted from 7.5×10^4 cells cultured under standard conditions were subjected to SDS-PAGE, blotted onto membranes and probed with anti-SHP1 or anti-β actin antibodies. (B) RT-PCR analysis of the *SHP1* gene. *SHP1* and β actin cDNAs were amplified by RT-PCR, then separated by 1.5% agarose gel electrophoresis. N.C. indicates the negative control, which did not contain cDNA. (C) MSP and unMSP assays of the *SHP1* promoter in hematopoietic cell lines and normal PBMCs. M and U indicate methylation and unmethylation specific PCR, respectively. (Koyama et al., 2003)

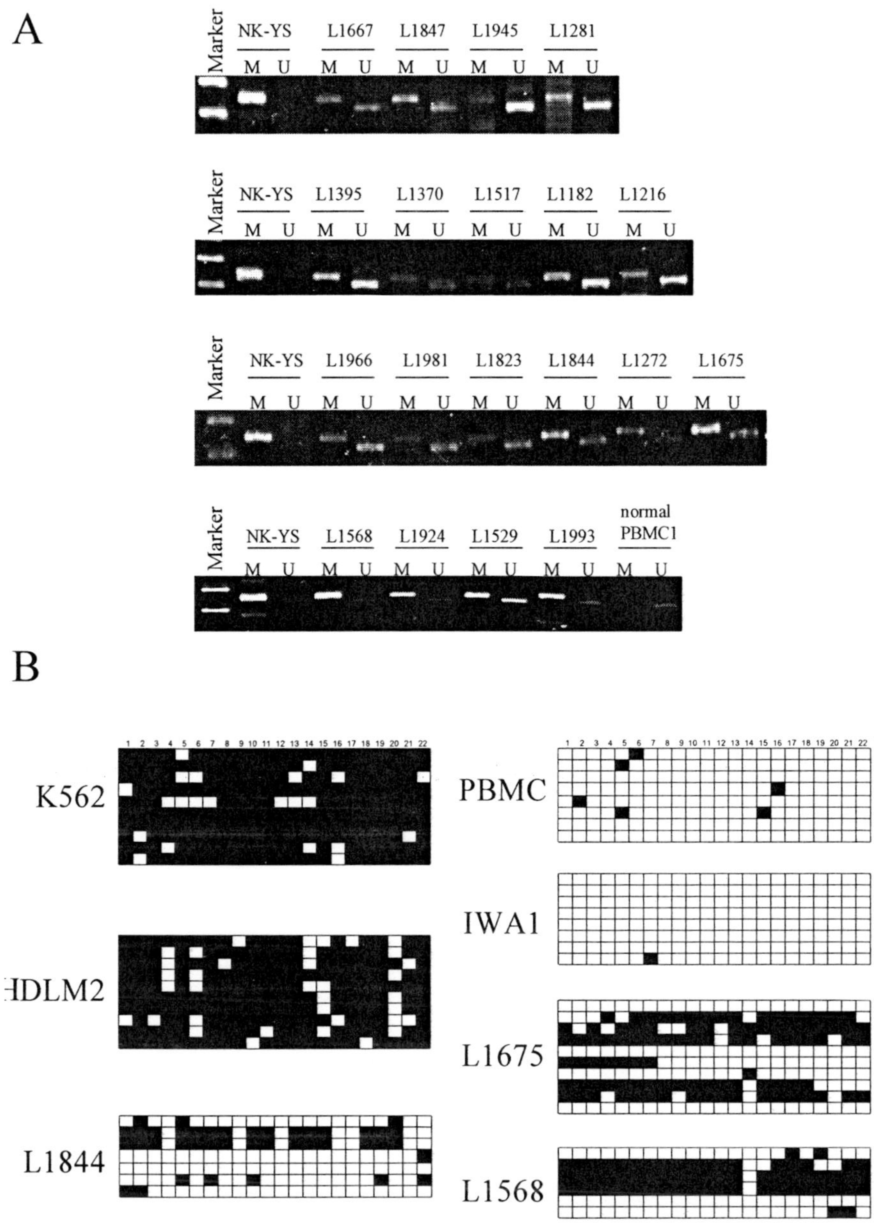

Figure 11. The *SHP1* gene expression and methylation analyses of patient specimens. (A) MSP and unMSP assays of the *SHP1* promoter in patient specimens. M and U indicate methylation and unmethylation specific PCR, respectively. DLBCL specimens, L1966, 1981, 1823 and 1844; MALT lymphoma specimens, L1667, 1847, 1272, 1675, 1568 and 1924; FL specimens, L1945, 1281, 1395, 1370, 1517, 1182, 1216, 1529 and 1993. (B) DNA methylation of the *SHP1* gene in cell lines and patient samples detected by DNA sequencing. Six to 10 clones of PCR products from bisulfite treated DNA were sequenced. Each row of squares represents a single cloned allele, and column represents each of the 22 possible methylated sites in the exon 1b promoter region of the *SHP1* gene (nucleotide numbers 7002–7520 of GenBank accession number AB079851). Filled and open squares represent methylated and unmethylated cytosines, respectively. (Koyama et al., 2003)

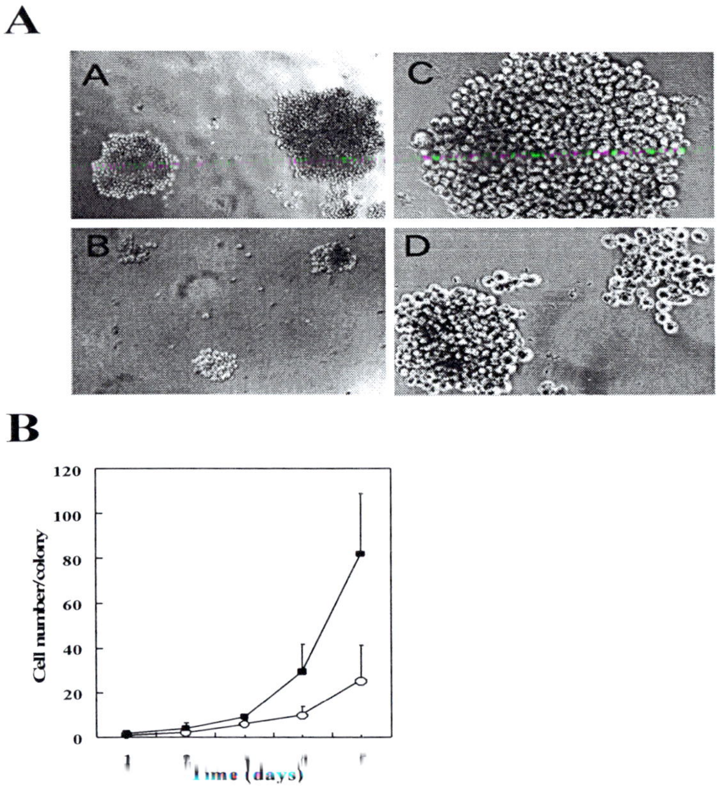

Figure 12. Growth suppression of K562 cells by exogenous SHP1. (A) K562 cells transfected with *SHP1* expression vector (b and d) or empty vector (a and c) were selected with G418 for 7 days, and observed with magnification of 40X (a and b) or 100X (c and d). (B) Cell number per colony was counted for the K562 cells transfected with *SHP1* expression vector (open circle) or empty vector (closed box). (Koyama et al., 2003)

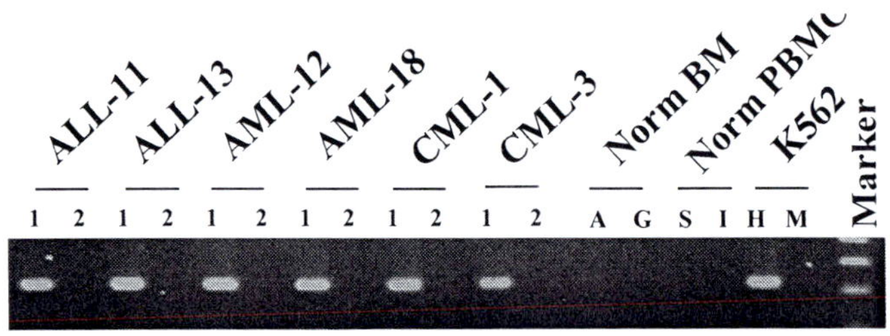

Figure 13. Loss of the *SHP1* promoter methylation according to complete remission. Methylation analysis by restriction enzyme-mediated PCR analysis. Patients or healthy donor DNAs were digested with methylation sensitive enzyme *Hpa*II and amplified the CpG island of the *SHP1* gene by PCR. 1, diagnostic PBMC's DNA from ALL or diagnostic bone marrow DNA from AML or CML patients. 2, bone marrow DNA obtained after hematological complete remission. Normal bone marrow DNA from two healthy volunteers (Norm BM) and normal PBMC DNA from two healthy volunteers (Norm PBMC) were used as control. K562 cell line, established from CML patient, was used as a positive control. H, *Hpa*II-digested DNA; M, *Msp*I-digested DNA. *Hpa*II is a methylation sensitive enzyme and *Msp*I is a methylation insensitive enzyme. Both enzymes recognize same DNA sequences. (Oka et al., 2002)

the unmethylated DNA for MSP analysis, bisulfite DNA sequencing were performed to determine the methylation of 22 CpG sites within the *SHP1* promoter that included the region examined by MSP and unMSP analyses (Figure 11). K562 and HDLM2 cell lines, which showed strong methylation signals by MSP analysis without signals by unMSP (Figure 10), were hypermethylated in all clones analyzed by bisulfite DNA sequencing (Figure 11B). It indicates that tumor cells, which totally methylated in the CpG island of the *SHP1* gene, do not express SHP1 protein (Figure 10 and 11). In contrast, almost no methylation was detected by bisulfite DNA sequencing (Figure 11) in all clones of IWA1 and normal PMBC, which showed no methylation signals by MSP and RE-PCR analyses and strong SHP1 protein expression by Western blot analysis. For the patient specimens such as L1844, L1675, and L1568, some clones showed hypermethylation, while others showed hypomethylation. These results suggested that the patient specimens contained both normal cells and blast cells that exhibited hypomethylation and hypermethylation, respectively (Figure 10) (Koyama et al., 2003).

SHP1 protein expression was recovered in B-cell lines after the treatment of demethylating reagent, 5-aza-2'-deoxycytidine, indicating CpG island methylation is actually contribute to the *SHP1* gene silencing (Oka et al., 2002; Koyama et al., 2003; Zhang et al, 2000). Transfection of the intact *SHP1* gene to the hematopoietic cultured cells, which show no expression of the *SHP1* gene, induced growth inhibition, indicating that gene silencing of the *SHP1* gene by aberrant methylation plays an important role to get the growth advantage of the malignant lymphoma/leukemia cells (Figure 12)(Bruecher-Encke et al., 2001; Koyama et al., 2003). The extra-ordinarily high frequency (75 to 100%) of CpG methylation of the *SHP1* gene in lymphoma/leukemia patient specimens indicates that the *SHP1* gene silencing is one of the critical events to the onset of malignant lymphomas/leukemias as well as it has important implications for the diagnostic or prognostic markers. From further investigation, methylation of the *SHP1* CpG island was revealed to be clearly correlated with the clinical stage of ALL, AML and CML patients; i.e., the *SHP1* gene was strongly methylated in the diagnostic PBMC of ALL patients and bone marrow of AML or CML patients, on the other hand no methylation of the *SHP1* promoter was detected in the hematological complete remission (Figure 13). These results indicate that methylation of the *SHP1* gene might be used for the monitoring or diagnosis of these leukemias as the clinical marker of expansion of malignant cells (Oka et al., 2002).

Genetic and Epigenetic Events in Lymphomagenesis/Leukemogenesis

We investigated loss of heterozygosity (LOH) with microsatellite markers located on the both sides of the *SHP-1* gene, in only the limited cases with hematological complete remission sample (Figure 14). Telomeric D12S356 marker showed LOH of 15 cases (79 %) in 19 informative cases and centromeric D12S336 marker showed 6 cases (38 %) of 16 informative cases by using diagnostic bone marrow samples and the peripheral blood samples obtained after hematological complete remission from ALL patients.

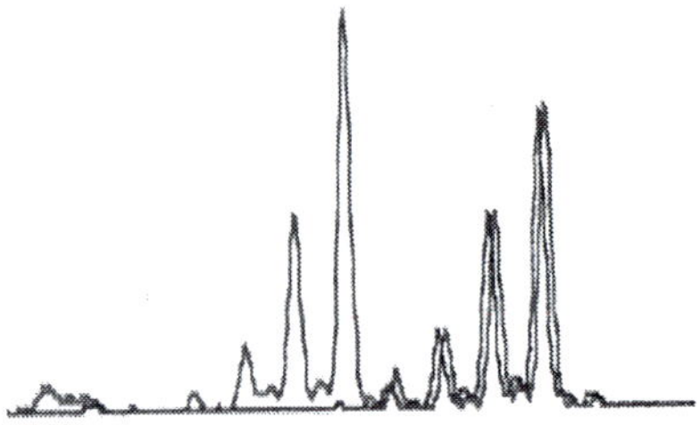

Microsatellite marker	*LOH*
D12S356	**15/19 (79 %)**
D12S336	**6/16 (38 %)**

Figure 14. Analyses of loss of heterozygosity (LOH) of ALL patients. Loss of heterozygosity (LOH) analysis was performed by using two microsatellite markers by which the *SHP-1* gene on chromosome 12p13 was surrounded. The diagnostic bone marrow sample (blue profile) and the peripheral blood sample obtained after hematological complete remission from ALL patient (red profile) showed the typical loss of heterozygosity with microsatellite marker D12S356. (Revised from Oka et al., 2002)

High frequent gene silencing of the *SHP1* gene with the epigenetic events like the DNA hypermethylation might be one of the critical event of lymphomagenesis and/or leukemogenesis as well as the genetic events such as mutational inactivation and allelic deletion to cause loss of hetrozygocity, eventually inducing completely abolishment of SHP1 protein expression. Loss of SHP1 protein expression will induce continuous cellular activation without down regulation of phosphorylation cascades such as JAK-STAT, PI3-kinase and MAP-kinase signaling pathways, causing stimulated and continued cell growth. Additional genetic changes and/or epigenetic events with histon modifications such as acetylation, phosphorylation and methylation causing chromosomal instability during the continuous cellular activation will promote the cells to the higher grade progression to the onset of leukemias and lymphomas (Figure 15).

Although SHP1 is primarily expressed in hematopoietic cells, over expression of SHP1 has been reported in breast cancer, ovarian cancer and prostate cancer (Mok et al., 1995; Yip et al., 2000; Zapata et al., 2002). Recent reports suggest that SHP1 may participate in the negative regulation of prostate cell proliferation by somatostatin and also in negative regulation of β-catenin signaling in epitherial cells (Zapata et al., 2002; Duchesne et al., 2003). Further investigation about these inhibitory functions of SHP1 would shed new insight on the involvement of somatostatin and/or β-catenin signaling pathways in the etiology of these cancers and could foster the development of new cancer therapies.

The importance of epigenetic gene scilencing in hematopoietic malignancies has been come forward understanding. Transcriptional repression of specific genes by CpG island methylation on the promoter region or histon deacetylation has been shown to play a relevant role in the pathogenesis of hematopoietic malignancies as well as cancers. Importantly, epigenetic gene silencing, as occurs through deacetylation of histons and/or DNA methylation, is a reversible process and can be modulated pharmacologically. It is quite contrast to the loss of gene function through mutations or deletions, which is also observed in hematopoietic malignancies and cancers. These genetic changes are not easily targeted pharmacological intervention, as the gene itself is permanently altered. Recently hypomethylation therapies with 5-aza-2'-deoxycytidine (Decitabine; SuperGen, Dublin,CA) have been reported the favorable results with AML, CML and myelodysplasia

(MDS)(Kantarjian et al., 2003; Issa et al., 2004). Inhibition of histondeacetylase (HDAC) by specific HDAC inhibitors such as trichostatin A and depsipeptide (FK228) has been shown to re-activate gene expression and to inhibit the growth and survival of tumor cells, and phase 1 trials have been reported in CLL and AML (Klisovic et al., 2003; Byrd et al., 2005). Although epigenetic therapy is hopeful, lots of problems remain to be elucidated – dose, patient selection, fundamental mechanism of the response, mechanism of resistance, safety and combination therapy for the clinical application of these epigenetic therapy.

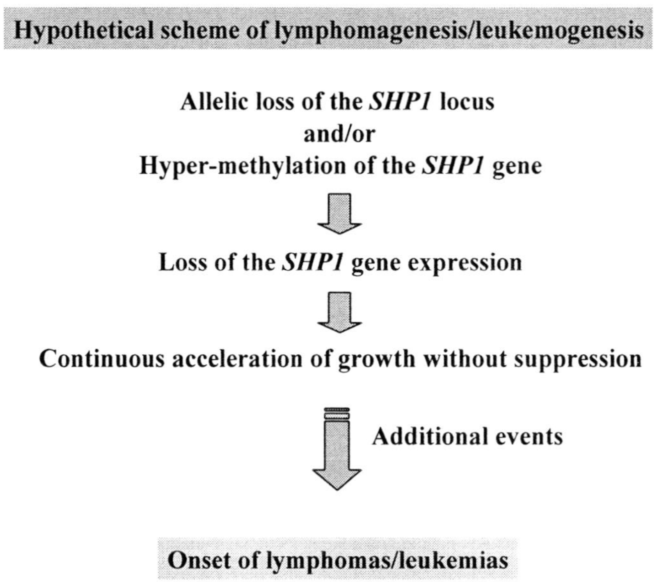

Figure 15. Hypothetical scheme of lymphomagenesis/leukemogenesis. The genetic events such as the *SHP1* locus allelic deletion and the epigenetic events such as hypermethylation of the CpG island of the *SHP1* gene induce completely abolishment of SHP1 protein expression. Loss of SHP1 protein expression will induce continuous cellular activation without down regulation of signaling cascades, causing stimulated and continued cell growth. Additional genetic or epigenetic events during such kinds of continuous cellular activation will promote the cells to the higher grade progression to the onset of leukemias and lymphomas.

Conclusion

From 75 to 100% of patient specimens of hematopoietic malignancies including leukemias and lymphomas were found to have gene silencing of the hematopoietic cell-specific protein-tyrosine phosphatase *SHP1* gene by CpG island methylation at the promoter region as well as in many hematopoietic cell lines, which is associated with the strong decrease of *SHP1* mRNA and loss of SHP1 protein expression. Genetic changes detected by loss of heterozygosity (LOH) with microsatellite markers near the *SHP1* gene were showed in 79% of informative ALL cases. The extra-ordinarily high frequency of CpG methylation of the *SHP1* gene in lymphoma/leukemia patient specimens indicates that the silencing of the *SHP1* gene with the genetic and/or epigenetic mechanisms is one of the critical events to the onset of malignant lymphomas/leukemias. The promoter CpG island methylation of the *SHP1* gene was clearly correlated with the clinical stage such as blast crisis and complete remission, indicating that methylation of the *SHP1* gene has important implications for the diagnostic or prognostic universal markers of hematopoietic malignancies and the target of gene therapy. These data strongly support the possibility that the *SHP1* gene is one of the tumor suppressor genes.

Aknowledgements

The authors gratefully acknowledge Dr. Maho Koyama (Department of Pediatrics, Okayama University Graduate School of Medicine and Dentistry), Yoko Nakatani, Takami Takabatake, Hiaki Satou, Masayuki Takano, Dr. Gui-Shan Jin, Dr. Shinichi Takada, Dr. Nobuya Ohara, Dr. Eisaku Kondo, Dr. Kiyoshi Takahashi, Dr. Tadaatu Akagi, Dr. Ichirou Murakami, (Department of Pathology, Okayama University Graduate School of Medicine and Dentistry) Dr. Yoichiro Ogama, (Department of Hematology, Oncology and Respiratory Medicine, Okayama University Graduate School of Medicine and Dentistry) and Dr. Kazuhiko Hayashi (Department of Molecular Pathology, Tottori University) for the contribution to this project. The authors also acknowledge Dr. Yoshinobu Matsuo (Fujisaki cell center, Hayashibara, Biomedical Laboratories, Okayama, Japan), Dr. E.C. Butcher (Department of Pathology, Stanford University, Stanford, CA), Dr Junjirou Tsuchiyama (Department of Medicine, Niigata University, Niigata, Japan) for provision of cell lines. The authors also acknowledge Mutumi Okabe, Yoshiko Sakamoto, Hiromi Nakamura, Yuki Onoda, and Miyuki Shiotani from the Department of Pathology, Okayama University Medical School, and Hiroshi Okamoto and Ai Tomosada from the Central Research Laboratory, Okayama University Medical School, for their excellent technical assistance. The authors also gratefully acknowledge Okayama University Hospital, National Okayama Medical Center, Fukuyama National Hospital, Okayama Saiseikai General Hospital, Toyama Prefectural Central Hospital, Chugoku Central Hospital and Okayama Red Cross General Hospital for generous supplying the patient specimens. This work was supported by the Ministry of Education, Culture, Sports, Science and Technology of Japan to T.O. (#12670161).

References

Alizadeh, A. A., Eisen, M. B., Izidore, D. R., Lossos, S., Rosenwald, A., Boldrick, J. C. Sabet, H., Tran, T., Yu, X., Powell, J. I., Yang, L., Marti, G. E., Moore, T., Hudson, J., Lu, L., Lewis, D. B., Tibshirani, R., Sherlock, G., Chan, W. C., Greiner, T. C., Weisenburger, D. D., Armitage, J. O., Warnke, R., Levy, R., Wilson, W., Grever, M. R., Byrd, J. C., Botstein, D., Brown, P. O. & Staudt. L. M. (2000). Distinct types of diffuse large B cell lymphoma identified by gene expression profiling. *Nature, 403,* 503-511.

Banville, D., Stocco, R. & Shen, S. H. (1995). Human protein tyrosine phosphatase 1C (PTPN6) gene structure: alternate promoter usage and exon skipping generate multiple transcripts. *Genomics, 27*, 165-173.

Beghini, A., Ripamonti, C.B., Peterlongo, P., Roversi, G., Cairoli, R., Morra, E. & Larizza, L. (2000). RNA hyperediting and alternative splicing of hematopoietic cell phosphatase (PTPN6) gene in acute myeloid leukemia. *Hum. Mol. Genet., 9*, 2297-2304.

Binstadt, B., Billadeau, D., Jevremovi, D., Williams, B. L., Fang, N., Yi, T., Koretzky, G. A., Abraham, R. T. & Leibson, P. J. (1998). SLP-76 is a direct substrate of SHP-1 recruited to killer cell inhibitory receptors. *J. Biol. Chem., 273,* 27518-27523.

Bittorf, T., Seiler, J., Zhang, Z., Jaster, R. & Brock, J. (1999). SHP1 protein tyrosine phosphatase negatively modulates erythroid differentiation and suppression of apoptosis in J2E erythroleukemic cells. *Biol. Chem., 380,* 1201-1209.

Blasioli, J., Paust, S. & Thomas, M. (1999). Definition of the sites of interaction between the protein tyrosine phosphatase SHP-1 and CD22. *J. Biol. Chem., 274,* 2303-2307.

Brizard, F., Dreyfus, B., Guilhot, F., Tanzer, J. & Brizard, A. (1997). 11q13 rearrangement in B cell chronic lymphocytic leukemia. *Leuk. Lymphoma, 25*, 539-43.

Brockdorff, J., Williams, S., Couture, C. & Mustelin, T. (1999). Dephosphorylation of ZAP70 and inhibition of T cell activation by activated SHP1. *Eur. J. Immunol., 29,* 2539-2550.

Brown, R. & Strathdee, G. (2002). Epigenomics and epigenetic therapy of cancer. *Trends Mol. Med. 8*, S43-48.

Bruecher-Encke, B., Griffin, J. D., Neel, B. G. & Lorenz, U. (2001). Role of the tyrosine phosphatase SHP-1 in K562 cell differentiation. *Leukemia, 15*, 1424-1432.

Byrd, J. C,, Marcucci, G., Parthun, M. R., Xiao, J. J., Klisovic, R. B., Moran, M., Lin, T. S., Liu, S., Sklenar, A. R., Davis, M. E., Lucas, D. M., Fischer, B., Shank, R., Tejaswi, S. L., Binkley, P., Wright, J., Chan, K. K. & Grever, M. R. (2005). A phase 1 and pharmacodynamic study of depsipeptide (FK228) in chronic lymphocytic leukemia and acute myeloid leukemia. *Blood, 105*, 959-967.

Carter, J. D., Neel, B. G. & Lorenz, U. (1999). The tyrosine phosphatase SHP-1 influences thymocyte selection by setting TCR signaling thresholds. *Int. Immunol., 11,* 1999-2014.

Chesi, M., Nardini, E., Brents, L. A., Schrock, E., Ried, T., Kuehl, W. M. & Bergsagel, P. L. (1997). Frequent translocation t(4;14)(p16.3;q32.3) in multiple myeloma is associated with increased expression and activating mutations of fibroblast growth factor receptor 3. *Nat. Genet.,16*, 260-264.

Chim, C. S., Fung, T. K., Cheung, W. C., Liang, R. & Kwong, Y. L. (2004a). SOCS1 and SHP1 hypermethylation in multiple myeloma: implications for epigenetic activation of the Jak/STAT pathway. *Blood, 103,* 4630-4635.

Chim, C. S., Wong, K.Y., Loong, F. & Srivastava, G. (2004b). SOCS1 and SHP1 hypermethylation in mantle cell lymphoma and follicular lymphoma: implications for epigenetic activation of the Jak/STAT pathway. *Leukemia, 18*, 356-358.

Cornall, R., Goodnow, C. & Cyster, J. (1999). Regulation of B cell antigen receptor signaling by the Lyn/CD22/SHP1 pathway. *Curr. Top. Microbiol. Immunol., 244,* 57-68.

Cyster, J. G. & Goodnow, C. C. (1995). Protein tyrosine phosphatase 1C negatively regulates antigen receptor signaling in B lymphocytes and determines thresholds for negative selection. *Immunity, 2*, 13-24.

Das, P. M. and Singal, R. (2004). DNA methylation and cancer. *J. Clin. Oncol., 22*, 4632-4642.

Delibrias, C. C., Floettmann, J. E., Rowe, M. & Fearon, D. T. (1997). Downregulated expression of SHP-1 in Burkitt lymphomas and germinal center B lymphocytes. *J. Exp. Med., 186*, 1575-1583.

Digilio, M. C., Pacileo, G., Sarkozy, A., Limongelli, G., Conti, E., Cerrato, F., Marino, B., Pizzuti, A., Calabro, R. & Dallapiccola, B. (2004). Familial aggregation of genetically heterogeneous hypertrophic cardiomyopathy: a boy with LEOPARD syndrome due to PTPN11 mutation and his nonsyndromic father lacking PTPN11 mutations. *Birth. Defects Res. A. Clin. Mol. Teratol., 70*, 95-98.

Duchesne, C., Charland, S., Asselin, C., Nahmias, C. & Rivard, N. (2003). Negative regulation of beta-catenin signaling by tyrosine phosphatase SHP-1 in intestinal epithelial cells. *J. Biol. Chem., 278*, 14274-14283.

Duggan, D.J., Bittner, M., Chen, Y., Meltzer, P. & Trent, J. M. (1999). Expression profiling using cDNA microarrays. *Nature Genet., 21,* 10-14.

Esteller, M. & Herman, J. G. (2002). Cancer as an epigenetic disease: DNA methylation and chromatin alterations in human tumours. *J. Pathol., 196,* 1-7.

Ferber, M. J., Eilers, P., Schuuring, E., Fenton, J. A., Fleuren, G. J., Kenter, G., Szuhai, K., Smith, D. I., Raap, A. K. & Brink, A. A. (2004). Positioning of cervical carcinoma and Burkitt lymphoma translocation breakpoints with respect to the human papillomavirus integration cluster in FRA8C at 8q24.13. *Cancer Genet. Cytogenet., 154*, 1-9.

Fusaki, N., Tomita, S., Wu, Y., Okamoto, N., Goitsuka, R., Kitamura, D. & Hozumi, N. (2000). BLNK is associated with the CD72/SHP-1/Grb2 complex in the WEHI231 cell line after membrane IgM cross-linking. *Eur. J. Immunol., 30,* 1326-1330.

Healy, J. & Goodnow, C. (1998). Positive versus negative signaling by lymphocyte antigen receptors. *Annu. Rev. Immunol., 16,* 645-670.

Issa, J. P, Garcia-Manero, G., Giles, F. J., Mannari, R., Thomas, D., Faderl, S., Bayar, E., Lyons, J., Rosenfeld, C. S., Cortes, J. & Kantarjian, H. M. (2004). Phase 1 study of low-dose prolonged exposure schedules of the hypomethylating agent 5-aza-2'-deoxycytidine (decitabine) in hematopoietic malignancies. *Blood, 103*, 1635-1640.

Jaffe, E.S., Chan, J. K. C., Su, I. J., Firizzera, G., Mori, S., Feller, A. C. & Ho, E.C. (1996). Report of the workshop on the nasal and related extranodal angiocentric T/natural killer cell lymphoma. *Am. J. Surg. Pathol., 20*, 103-111.

Jin, Y. J., Yu, C. L. & Burakoff, S. J. (1999). Human 70-kDa SHP-1L differs from 68-kDa SHP-1 in its C-terminal structure and catalytic activity. *J. Biol. Chem., 274*, 28301-28307.

Kantarjian, H. M., O'Brien, S., Cortes, J., Giles, F. J., Faderl, S., Issa, J. P., Garcia-Manero, G., Rios, M. B., Shan, J., Andreeff, M., Keating, M. & Talpaz M. (2003). Results of decitabine (5-aza-2'deoxycytidine) therapy in 130 patients with chronic myelogenous leukemia. *Cancer, 98*, 522-528.

Khoury, J. D., Rassidakis, G. Z., Medeiros, L. J., Amin, H. M. & Lai, R. (2004). Methylation of SHP1 gene and loss of SHP1 protein expression are frequent in systemic anaplastic large cell lymphoma. *Blood, 104*, 1580-1581.

Klisovic, M. I., Maghraby, E. A., Parthun, M. R., Guimond, M., Sklenar, A. R., Whitman, S. P., Chan, K. K., Murphy, T., Anon, J., Archer, K. J., Rush, L. J., Plass, C., Grever, M.R., Byrd, J. C. & Marcucci, G. (2003). Depsipeptide (FR 901228) promotes histone acetylation, gene transcription, apoptosis and its activity is enhanced by DNA methyltransferase inhibitors in AML1/ETO-positive leukemic cells. *Leukemia, 17*, 350-358.

Komatsu, H., Yoshida, K., Seto, M., Iida, S., Aikawa, T., Ueda, R. & Mikuni, C. (1993). Overexpression of PRAD1 in a mantle zone lymphoma patient with a t(11;22)(q13;q11) translocation. *Br. J. Haematol., 85*, 427-429.

Kononen, J., Bubendorf, L., Kllioniemi, A. Barlund, M., Schraml, P., Leighton, S., Torhorst, J., Mihatsch, M. J. & Sauter, G. (1998). Tissue microarrays for high-throughput molecular profiling of tumor specimens. *Nature Med., 4,* 844-847.

Kossev, P. M., Raghunath, P. N., Bagg, A., Schuster, S., Tomaszewski, J. E. & Wasik, M. A. (2001). SHP-1 expression by malignant small B-cell lymphomas reflects the maturation stage of their normal B-cell counterparts. *Am. J. Surg. Pathol., 25*, 949-955.

Koyama, M., Oka, T., Ouchida, M., Nakatani, Y., Nishiuchi, R., Yoshino, T., Hayashi, K., Akagi, T. & Seino, Y. (2003). Activated proliferation of B-cell lymphomas/leukemias with the SHP1 gene silencing by aberrant CpG methylation. *Lab. Invest. 83*, 1849-1858.

Ladetto, M., Mantoan, B., Ricca, I., Astolfi, M., Drandi, D., Compagno, M., Vallet, S., dell'Aquila, M., Alfarano, A., Rossatto, P., Rocci, A., Vitolo, U., Corradini, P., Boccadoro, M. & Tarella, C. (2003). Recurrence of Bcl-2/IgH polymerase chain reaction positivity following a prolonged molecular remission can be unrelated to the original follicular lymphoma clone. *Exp. Hematol., 31,* 784-788.

Lander, E.S., et al. (2001). Initial sequencing and analysis of the human genome. *Nature, 409*, 860-921.

Legius, E., Schrander-Stumpel, C., Schollen, E., Pulles-Heintzberger, C., Gewillig, M., Fryns, J. P. (2002). PTPN11 mutations in LEOPARD syndrome. *J. Med. Genet., 39*, 571-574.

Leon, F., Cespon, C., Franco, A., Lombardia, M., Roldan, E., Escribano, L., Harto, A., Gonzalez-Porque, P. (2002). Roy, G. SHP-1 expression in peripheral T cells from patients with Sezary syndrome and in the T cell line HUT-78: implications in JAK3-mediated signaling. *Leukemia, 16*, 1470-1477.

Loh, M. L., Reynolds, M. G., Vattikuti, S., Gerbing, R. B., Alonzo, T. A., Carlson, E., Cheng, J. W., Lee, C. M., Lange, B. J. & Meshinchi, S. (2004). Children's Cancer Group. PTPN11 mutations in pediatric patients with acute myeloid leukemia: results from the Children's Cancer Group. *Leukemia, 18,* 1831-1834.

Lorenz, U., Bergemann, A. D., Steinberg, H. N., Flanagan, J. G., Li, X., Galli, S.J. & Neel, B. G. (1996a). Genetic analysis reveals cell type-specific regulation of receptor tyrosine kinase c-Kit by the protein tyrosine phosphatase SHP1. *J. Exp. Med., 184,* 1111-1126.

Lorenz, U., Ravichandran, K. S., Burakoff, S. J. & Neel, B. G. (1996b). Lack of SHPTP1 results in src-family kinase hyperactivation and thymocyte hyperresponsiveness. *Proc. Natl. Acad. Sci. U S A, 93*, 9624-9629.

Ma, X. Z., Jin, T., Sakac, D., Fahim, S., Zhang, X., Katsman, Y., Bali, M. & Branch, D. R. (2003). Abnormal splicing of SHP-1 protein tyrosine phosphatase in human T cells. Implications for lymphomagenesis. *Exp. Hematol., 31*, 131-142.

Magrath, I. (1995). The pathogenesis of Burkitt's lymphoma. *Adv. Cancer Res., 55*, 133-270.

Martinelli, G., Tosi, P., Ottaviani, E., Soverini, S. & Tura, S. (2001). Molecular therapy for multiple myeloma. *Haematologica, 86*, 908-917.

Matsushita, M., Tsuchiya, N., Oka, T., Yamane, A. & Tokunaga, K. (1999). New variations of human SHP-1. *Immunogenetics, 49*, 577-579.

Migone, T. S., Cacalano, N. A., Taylor, N., Yi, T., Waldmann, T. A. & Johnston, J. A. (1998). Recruitment of SH2-containing protein tyrosine phosphatase SHP-1 to the interleukin 2 receptor; loss of SHP-1 expression in human T-lymphotropic virus type I-transformed T cells. *Proc. Natl. Acad. Sci. USA, 95,* 3845-3850.

Mizuno, K., Katagiri, T., Hasegawa, K., Ogimoto, M. & Yakura, H. (1996). Hematopoietic cell phosphatase, SHP-1, is constitutively associated with the SH2 domain-containing leukocyte protein, SLP-76, in B cells. *J. Exp. Med., 184,* 457-463.

Moch, H., Schraml, P., Burbendorf, L., Mirlacher, M., Kononen, J., Gasser, T., Mihatsch, M. J., Kallioniemi, O.P. & Sauter, G. (1999). High-throughput tissue microarray analysis to evaluate genes uncovered by cDNA microarray screening in renal cell carcinoma. *Amer. J. Pathol., 154*, 981-986.

Mok, S. C., Kwok, T. T., Berkowitz, R. S., Barrett, A. J. & Tsui, F. W. (1995). Overexpression of the protein tyrosine phosphatase, nonreceptor type 6 (PTPN6), in human epithelial ovarian cancer. *Gynecol. Oncol., 57*, 299-303.

Nakanishi, T., Oka, T. & Akagi, T. (2001). Recent advances in DNA microarrays. *Acta Med. Okayama, 55,* 319-328.

Oka, T., Yoshino, T., Hayashi, K., Ohara, N., Nakanishi, T., Yamaai, Y., Hiraki, A., Aoki Sogawa, C., Kondo, E., Teramoto, N., Takahashi, K., Tsuchiyama, J. & Akagi, T. (2001). Reduction of hematopoietic cell-specific tyrosine phosphatase SHP-1 gene expression in natural killer cell lymphoma and various types of lymphomas/leukemias: combination analysis with cDNA expression array and tissue microarray. *Am. Journal of Pathology, 159*,1495-1505.

Oka, T., Ouchida, M., Koyama, M., Ogama, Y., Takada, S., Nakatani, Y., Tanaka, T., Yoshino, T., Hayashi, K., Ohara, N., Kondo, E., Takahashi, K., Tsuchiyama, J., Tanimoto, M., Shimizu, K. & Akagi, T. (2002). Gene silencing of the tyrosine phosphatase SHP1 gene by aberrant methylation in leukemias/lymphomas. *Cancer Res, 62, 6*390-6394.

Ohno, H. (2004). Pathogenetic role of BCL6 translocation in B-cell non-Hodgkin's lymphoma. *Histol. Histopathol., 19,* 637-650.

Pani, G., Kozlowski, M., Cambier, J.C., Mills, G. B. & Siminovitch, K. A. (1995). Identification of the tyrosine phosphatase PTP1C as a B cell antigen receptor-associated protein involved in the regulation of B cell signaling. *J. Exp. Med., 181*, 2077-2084.

Pani, G., Fischer, K. D., Mlinaric-Rascan, I. & Siminovitch, K. A. (1996). Signaling capacity of the T cell antigen receptor is negatively regulated by the PTP1C tyrosine phosphatase. *J. Exp. Med., 184*, 839-852.

Pei, D., Lorenz, U., Klingmuller, U., Neel, B.G. & Walsh, C.T. (1994). Intramolecular regulation of protein tyrosine phosphatase SH-PTP1: a new function for Src homology 2 domains. *Biochemistry, 33,* 15483-15493.

Plutzky, J., Neel, B. G., Rosenberg, R. D., Eddy, R. L., Byers, M. G., Jani-Sait, S. & Shows, T. B. (1992). Chromosomal localization of an SH2-containing tyrosine phosphatase (PTPN6). *Genomics, 13*, 869-872.

Poetsh, M., Weber-Matthiesen, K., Plendl, H. J., Grote, W. & Schlegelberger, B. (1996). Detection of the t(14;18) chromosomal translocation by interphase cytogenetics with yeast-artificial-chromosome probes on follicular lymphoma and nonneoplastic lymphoproliferation. *J. Clin. Oncol.,14*, 963-969.

Razin, A. & Cedar, H. (1991). DNA methylation and gene expression. *Microbiol. Rev., 55*, 451-458.

Schraml, P., Kononen, J., Burbendorf, L., Moch, H., Bissig, H., Nocito, A., Mihatsch, M. J., Kallioniemi, O. P. & Sauter, G. (1999). Tissue microarrays for gene amplification surveys in many different tumor types. *Clin. Cancer Res., 5*, 1966-1975.

Shen, S. H., Bastien, L., Posner, B. I. & Chretien, P. (1991). A protein-tyrosine phosphatase with sequence similarity to the SH2 domain of the protein-tyrosine kinases. *Nature, 352,* 736-739.

Shultz, L. D., Schweitzer, P. A., Rajan, T. V., Yi, T., Ihle, J. N., Matthews, R. J., Thomas, M. L. & Beier, D. R. (1993). Mutations at the murine motheaten locus are within the hematopoietic cell protein-tyrosine phosphatase (Hcph) gene. *Cell, 73,* 1445-1454.

Tartaglia, M., Mehler, E. L., Goldberg, R., Zampino, G., Brunner, H. G., Kremer, H., van der Burgt, I., Crosby, A. H., Ion, A., Jeffery, S., Kalidas, K., Patton, M.A., Kucherlapati, R.S. & Gelb, B.D. (2001). Mutations in *PTPN11,* encoding the protein tyrosine phosphatase SHP -2, cause Noonan syndrome. *Nat. Genet., 29*, 465–468.

Tsui, H. W., Siminovitch, K. A., de Souza, L. & Tsui, F. W. L. (1993). Motheaten and viable motheaten mice have mutations in the haematopoietic cell phosphatase gene. *Nature Genet., 4,* 124-129.

Tsui, H.W., Hasselblatt, K., Martin, A., Mok, S. C. & Tsui, F. W. (2002). Molecular mechanisms underlying SHP-1 gene expression. *Eur. J. Biochem., 269*, 3057-3064.

Venter, J.C., et al. (2001). The sequence of the human genome. *Science, 291,* 1304-1351.

Yang, J., Liang, X., Niu, T., Meng, W., Zhao, Z. & Zhou, G. W. (1998). Crystal structure of the catalytic domain of protein-tyrosine phosphatase SHP-1. *J. Biol. Chem., 273*, 28199-28207.

Yang, J., Cheng, Z., Niu, T., Liang, X., Zhao, Z. J. & Zhou, G. W. (2000). Structural basis for substrate specificity of protein-tyrosine phosphatase SHP-1. *J. Biol. Chem., 275*, 4066-4071.

Yang, J., Liu, L., He, D., Song, X., Liang, X., Zhao, Z. J. & Zhou, G. W. (2003). Crystal structure of human protein-tyrosine phosphatase SHP-1. *J. Biol. Chem., 278*, 6516-6520.

Yang, W., Tabrizi, M., Berrada, K. & Yi, T. (1998). SHP-1 phosphatase C-terminus interacts with novel substrates p32/p30 during erythropoietin and interleukin-3 mitogenic responses. *Blood, 91,* 3746-3755.

Ye, B. H., Chaganti, S., Chang, C. C., Niu, H., Corradini, P., Chaganti, R. S. & Dalla-Favera, R. (1995). Chromosomal translocations cause deregulated BCL6 expression by promoter substitution in B cell lymphoma. *EMBO J., 14*, 6209-6217.

Yip, S. S., Crew, A. J., Gee, J. M., Hui, R., Blamey, R. W., Robertson, J. F., Nicholson, R. I., Sutherland, R. L. & Daly, R. J. (2000). Up-regulation of the protein tyrosine phosphatase SHP-1 in human breast cancer and correlation with GRB2 expression. *Int. J. Cancer., 88*, 363-368.

Yoshida, R., Hasegawa, T., Hasegawa, Y., Nagai, T., Kinoshita, E., Tanaka, Y., Kanegane, H., Ohyama, K., Onishi, T., Hanew, K., Okuyama, T., Horikawa, R., Tanaka, T. & Ogata, T. (2004). Protein-tyrosine phosphatase, nonreceptor type 11 mutation analysis and clinical assessment in 45 patients with Noonan syndrome. *J. Clin. Endocrinol. Metab., 89*, 3359-3364.

Wang, K., Gan, L., Jeffery, E., Gayle, M., Gown, A. M., Skelly, M., Nelson, P.S., Ng, W. V. & Schummer, M. (1999). Monitoring gene expression profile changes in ovarian carcinomas using cDNA microarray. *Gene, 229*, 101-108.

Wang, L. L., Blasioli, J., Plas, D., Thomas, M. & Yokoyama, W. (1999). Specificity of the SH2 domains of SHP-1 in the interaction with the immunoreceptor tyrosine-based inhibitory motif-bearing receptor gp49B. *J. Immunol., 162,* 1318-1322.

Wu, Y., Nadler, M., Brennan, L., Gish, G. D., Timms, J. F., Fusaki, N., Jongstra, B., J., Tada, N., Pawson, T., Wither, J., Neel, B. G. & Hozumi, N. (1998). The B-cell transmembrane protein CD72 binds to and is an in vivo substrate of the protein tyrosine phosphatase SHP-1. *Curr. Biol., 8,* 1009-1017.

Zapata, P. D., Ropero, R. M., Valencia, A. M., Buscail, L., Lopez, J. I., Martin-Orozco, R. M., Prieto, J. C., Angulo, J., Susini, C., Lopez-Ruiz, P. & Colas, B. (2002). Autocrine regulation of human prostate carcinoma cell proliferation by somatostatin through the modulation of the SH2 domain containing protein tyrosine phosphatase (SHP)-1. *J. Clin. Endocrinol. Metab., 87*, 915-926.

Zhang, L., Zhou, W., Velculescu, V. E., Kern, S. E., Hruban, R. H., Hamilton, S. R., Vogelstein, B. & Kinzler. K. W. (1997). Gene expression profiles in normal and cancer cells. *Science, 276*, 1268-1272.

Zhang, Q., Lee, B., Korecka, M., Li, G., Weyland, C., Eck, S., Gessain, A., Arima, N., Lessin, S. R., Shaw, L. M., Luger, S., Kamoun, M. & Wasik, M. A. (1999). Differences in phosphorylation of the IL-2R associated JAK/STAT proteins between HTLV-I(+), IL-2-independent and IL-2-dependent cell lines and uncultured leukemic cells from patients with adult T-cell lymphoma/leukemia. *Leuk. Res., 23,* 373-384.

Zhang, Q., Raghunath, P. N., Vonderheid, E., Odum, N. & Wasik, M.A. (2000). Lack of phosphotyrosine phosphatase SHP-1 expression in malignant T-cell lymphoma cells results from methylation of the SHP-1 promoter. *Am. J. Pathol., 157,* 1137-1146.

In: Gene Silencing: New Research
Editor: Grace W. Redberry, pp. 33-80
ISBN 1-59454-832-3

Chapter II

Age Management Treatments which Target Silenced Genes

Stanislaw R. Burzynski
Burzynski Clinic
Houston, Texas U.S.A.

Abstract

Successful completion of the human genome project increased interest in mechanisms controlling gene expression. Data presented by many research groups revealed silencing of genes in aging. Methylation in promoters regions, histone modification and RNAi mechanisms silence the genes which slow down the aging. Global DNA demethylation increases genomic instability, amplification of oncogenes and silencing of the genes through the RNAi mechanism, which accelerates the process. In aging, there is a decreased expression of tumor suppressors and genes involved in detoxification and cholesterol metabolism but, to the contrary, there is an increased expression of the oncogenes and the genes instrumental in chronic inflammation, stress response and silencing. Gene silencing affects crucial processes, such as cell cycle, metabolism, programmed cell death, nuclear transport, endogenous clocks and many others. Normal balance in gene expression, unlike mutations, can be restored by activation of silenced genes through the use of agents which decrease methylation of promoters, induce acetylation of histones and normalize global methylation of DNA. Small organic molecules, phenylacetylglutamine (PG) and phenylacetylisoglutamine (isoPG), increase expression of tumor suppressor genes *PTEN* and *MAD* and down regulate oncogenes *AKT2* and *MYCC*. Cosmetic cream of PG and isoPG was studied in clinical trial, which provided evidence of its antiaging efficacy on the skin. The reversal of gene silencing and down regulation of oncogenes are attractive targets in age management.

Introduction

The establishment of the sequence of the genome revealed that less than 2% of human genome corresponds to genes which code proteins and only approximately 1/10 of them are active in the adult individual. This means that approximately 90% of human genes are silenced. It has been recognized that silencing and activation of the genes is a dynamic process which continues during the entire life. Starting from almost complete demethylation of the genome at the beginning of the life, the activity of various groups of the genes is turned off and the main reason is that their function is no longer needed (Li, 2002). The activity of many genes is necessary during dramatic development in embryonal and fetal life but after the organogenesis, many important genes are silenced. Dramatic changes in gene activity occur soon after the childbirth and the additional genes, such as hemoglobin F, are silenced at that time (Karlsson and Nienhuis, 1985). Perhaps the ideal set of active and silenced genes occurs in the body of a 25-year-old human, but soon after that many important genes are silenced. According to the author's theory of aging, the aging process involves silencing of the genes and contributes to many other chronic pathological processes; among them cancer (Burzynski, 2003). While some of the genes are silenced during aging, the others increase the activity due to a disturbed balance with inactive genes and also due to the effect of numerous endogenous and external factors. During the aging process, important tumor suppressor genes are silenced and the oncogenes become more active. Extremely complicated changes in the activity of the genes require an additional biochemical system named epigenome which consists of molecular switches regulating the genome (Feinberg and Tycko, 2004).

This chapter summarizes molecular mechanisms of gene silencing and briefly describes the agents developed by our team which may restore normal gene expression and activate silenced genes. It describes gene silencing theory of aging and changes in gene expression in skin which is the largest organ of the body. Finally, the chapter describes preliminary attempts to use the principle of activation of silenced genes in age management.

Molecular Mechanisms of Gene Silencing

DNA Methylation

Background

DNA methylation is one of the most important chemical processes involved in regulation of gene expression (Ordway and Curran, 2002; Robertson, 2002; Esteller, 2005). In the human body, most DNA methylation occurs in cytosine followed by guanosine (CpG sites). The regions of DNA rich in CpG dinucleotides are named "CpG islands" (CGIs). They are usually found in gene promoter regions, but they are also located in exons, introns and other positions of the gene (Takai and Jones, 2002). DNA has unequal distribution of CpG dinucleotides, including large areas where they are not present. Outside genes, methylated cytosine is found in repetitive sequences, retrotransposons and endogenous retroviral elements (Yoder et al., 1997). In the silenced genes, promoter associated CGIs are generally methylated and it is assumed that such methylation is essential for maintenance of gene inactivity (Issa, 2004). Analysis of silenced and non-silenced genes revealed that silencing is associated with methylation of a small "core" region of the promoter extending to the transcription start (Kaneda et al., 2002; Hagihara et al., 2004; Ushijima, 2005). In aging and cancer, usually there is both global hypomethylation and hypermethylation of CGIs in promoter regions of tumor suppressor genes. Histone modification, as suggested by earlier studies, is secondary to DNA methylation, but according to more recent research it may occur before DNA methylation (Tamaru and Selker, 2001).

Biochemistry of Methylation

Methylation Enzymes

The reaction of methylation is mediated by ternary enzyme complex consisting of methyltransferases (DNMTs), methionine adenosyltransferases (MATs), and S-adenosyl-homocysteine hydrolases (SAHHs). The three isozymes of MAT and SAHH have been identified (Liau and Burzynski, 1986; Liau et al., 1992). Cancer cells have different isozymes: MAT-LT and SAHH-LT. The abnormal MAT-LT was shown to be associated with cancer specific proteins. Such enzymes form a stable ternary complex, in contrast with normal enzymes which disassociate very easily. This allows cancer cells to keep methylation enzymes permanently in the active state, providing a locked-in mechanism to enable them to drive neoplastic cell proliferation (Liau and Burzynski, 1986; Liau et al., 1992). MAT enzymes carry methyl groups from *S*-adenosylmethionine to DNA. The attachment of methyl groups to DNA is catalyzed by DNMTs. SAHH enzymes catalyze hydrolysis of *S*-adenosyl-homocysteine to homocysteine and adenosine. At present most research in this area concentrates on DNMTs. Currently there are three groups of DNMTs (Robertson, 2002). Group 1 includes DNMT1, DNMT1b, DNMT1o and DNMT1p. Group 3 consists of DNMT3A, DNMT3b and DNMT3L. At present, there is only one DNMT2. Demethylation enzymes include 5-methylcytosine glycosylase (Fremont et al., 1997) and methyl-CpG binding domain 2 demethylase (HBD2b) (Bhattacharya et al., 1999). The mechanism of action of 5- methylcytosine glycosylase is based on elimination of methylcytosine from the deoxyribose of DNA which is followed by the addition of new cytosine to the DNA chain by DNA repair enzymes. MBD2b hydrolyses 5-methylcytosine to cytosine, but these results were not yet reproduced (Ng et al., 1999).

Methyl Binding Proteins

An additional process of transcriptional repression involves methyl-CpG-binding proteins (MBPs) and recruitment of histone deacetylases (HDACs) (Ordway and Curran, 2002). Methylated CpG reacts with MBP and mediates turning of transcription. Five MBPs have been initially identified: MeCP1, MeCP2, MBD1, MBD2 and MBD3 (Prokhortchouk and Hendrich, 2002). In addition, MBD4 is involved in DNA repair and Kaiso binds through zinc finger motif, instead through the methyl CpG binding domain (MBD) used by other MBPs. (Prokhortchouk et al., 2001). MBD is located at N-terminus at MBP, whereas C-terminal domain represses transcription. MeCP2, MBD1, MBD2, and Kaiso bind to methylated CpGs and recruit HDACs which causes condensation of chromatin and blocks transcription (Prokhortchouk and Heindrich, 2002). HDAC inhibitors including trichostatin A block the action of MBPs.

Hypermethylation

Numerous genes are silenced in aging and cancer through hypermethylation of promoters regions. It is assumed that the primary cause of increased methylation in cancer is methylation in normal tissue associated with aging (Issa et al., 1994; Issa, 2002). Aging is associated with cell renewal requiring repeated mobilization of stem cells. The extension of DNA methylation is linked to *de novo* methylation and replication (Valicescu et al., 2002). According to Turker, the process starts during embryonal life at the "methylation centers" which are typically repetitive elements such as retrotransposons (Turker, 2002). DMMT's are attracted to specific sequences of these elements (see also section *RNAi Mechanisms).* In aging process methylation has a tendency to spread toward promoters and is blocked by CGIs (Issa, 2004). Unmethylated DNA is probably protected against spread of methylation by transcription factors and other unidentified proteins. Numerous factors may disrupt such protection and promote the spread of methylation to CGIs and gene silencing. They include decreased expression of protective genes (silenced in aging), mutations which activate DNMTs, diet, chronic inflammation, carcinogens and viral infection (Issa, 2004; Issa et al., 1994; 2001; Waterland and Jirtle, 2003; Kang et al., 2002). Loss of activity of a single gene may affect many others, leading to CGI methylation phenotype (CIMP) and cancer (Issa, 2004). Silencing of genes in cancer has been described in numerous articles and includes *p15*^{INK4b} *(p15), p16*^{INK4a} *(p16), retinoblastoma (RB), p14*^{ARF} *(p14), p21*$^{Cip1/Waf1}$ *(p21)* and *TP53* involved in cell cycle control (Herman et al., 1995; Villuendas et al., 1998; Garcia et al., 2002; Gonzales-Gomez et al., 2003; Leone et al., 2002). Additional genes, which are silenced through hypermethylation are involved in DNA repair, including *BRCA* and *MGTM*, apoptosis (*DAPK*) and detoxification (*GSTP1*) (Dobrovic and Simpfendorfer, 1997; Harden et al.; 2003; Hegi et al., 2004; Lee et al., 1994).

Hypomethylation

Global Hypomethylation

In aging, there is a decrease of total genomic methylation (Hoal-van Helden and van Helden, 1989), but there is an increase of methylation of CGIs (Issa et al., 1994; Ahuja et al., 1998). In cancer, these changes are more pronounced with a genome-wide hypomethylation and marked increase of methylation of CGIs. Global hypomethylation in neoplasia was found in every type of cells studied, both benign and malignant (Feinberg and Tycko, 2004; Gama-Sosa et al., 1983; Goelz et al., 1985). The evidence was presented that hypomethylation was increasing activity of oncogenes, including *HRAS* and *Cyclin D2* and genes associated with proliferations, such as *14-3-3*σ (Feinberg and Tycko, 2004; Oshimo et al., 2003; Akiyama et al., 2003; Sato et al., 2003). Decreased global methylation leads to genomic instability Hypomethylation in pericentromeric satellite sequences in breast and ovarian cancers and Wilms tumor leads to chromosomal translocations with DNA breakpoints in chromosomes 1 and 16 (Qu et al., 1999; Yeh et al., 2002). The global hypomethylation in brain tumors is increasing with a higher grade of malignancy (Ehrlich, 2002). Hypomethylation of multidrug-resistance gene *MDR1* increases resistance to cytotoxic chemotherapy (Nakayama et al., 1998).

Hypomethylation and Chromatin Remodeling Complex

Global hypomethylation is linked to the activity of SWI/SNF complex which includes INI1 (SNF5) subunit (described in detail in a Chromatin Remodeling section). Patients with atypical teratoid/rhabdoid tumors (AT/RTs) who have mutation of *INI1* gene have extensive global demethylation (Feinberg and Tycko, 2004). Patients with α-thalassemia, myelodysplasia disorder (ATRX) have decreased methylation of ribosomal DNA repeats (Gibbons et al., 2000). Mutated product of *ATRX* gene encodes SNF2 group helicase. Another member of SNF2 family, Lsh is also required for maintenance of global methylation (Fan et al., 2003).

Hypomethylation of Mobile Elements

An important event is hypomethylation of latent retrotransposons (Alves et al., 1996; Singer et al., 1993), especially long interspersed nuclear elements (LINES) which are common mobile DNA elements. This leads to transcriptional activation and occurs in various types of cancer, including bladder carcinoma (Jurgens et al., 1996). It causes changes in the expression of the adjacent genes such as *APC* and *MYCC* in colon and breast cancers (Costello and Plass, 2001). . Hypomethylation of retrotransposons promotes chromosomal rearrangements in colon cancer (Suter et al., 2004).

RNAi Mechanisms

Global hypomethylation of DNA increases genomic instability and leads to chromosomal abnormalities and silencing of the genes through RNAi. "Junk DNA" consists mainly of retrotransposons and their remnants (long terminal repeats-LTRs) which can generate small RNAs essential for regulating cell differentiation and centromere functions. In fission yeast, RNA-dependent RNA polymerase (Rdpl) generates double-stranded RNA (dsRNA). Dicer cleaves dsRNA to produce siRNA which attaches to DNA of LTR and forms a region of RNA-DNA sequence identity. DNMT recognizes a RNA-DNA hybrid as a substrate and this leads to recruitment of histone modifying enzymes and methylation of histone H3 lysine 9 (H3K9). Heterochromatin protein 1 (HP1) is recruited and spreads formation of compact chromatin to nearby genes. RNAi-based mechanisms contribute to the control of cellular differentiation by acting on LTRs of the retrotransposons and presumably initiating the spread of silent heterochromatin into nearby genes (Schramke and Allshire, 2003).

Chromatin Remodeling

Background

The processes which are associated with methylation of DNA are chromatin remodeling and histone acetylation. Hypermethylated CpGs only have functional importance if they effect chromatin changes. An active gene must be selectively presented from the chromosome in order to gain access to RNA polymerase, but the inactive gene remains coiled and is not transcribed. Linker DNA of 20-60 bp connects nucleosomes and forms 10 nm fiber – "beads-on-a-string" which corresponds to active gene. Compaction of chromatin into 30 nm fiber silences the gene because it reduces the accessibility of DNA (Horn and Peterson, 2002). Research of Dorigo et al. indicated that a compacted chromatin fiber has a "two-start helix" structure. In this structure "beads-on-a-string" nucleosomes traverse back and forth between two adjacent helical columns (Dorigo et al., 2004). The chromatin structure is modified by chromatin remodeling enzymes (Fry and Peterson, 2001) which change nucleosome position and enzymes that modify histone tails (described in Histone Modification section) (Roth et al., 2001; Jenuwein and Allis, 2001). Transcription factors and acetyltransferases (HATs) are assembled at the transcriptional sites (Chan and La Thangue, 2001; Bulger et al., 2002). Histone modifications mark domains of the genes which are regulated independently from the others (Noma et al., 2001; Litt et al., 2001). It is suggested that a system of "landing platforms" exists for chromatin remodeling factors including activators or repressors, depending on the genomic site. One of these regulatory factors is a special AT-rich binding protein 1 (SATB1) (Cai et al., 2003). It is proposed that this factor brings specialized DNA sequences to its "cage-like network" and regulates tissue-specific gene expression through its effect on higher order chromatin structure (Cai et al., 2003). The data presented indicate that SATB1 regulates induction of MYCC in thymocytes. Two classes of proteins are involved in opening and closing of the chromatin structure; trithorax (trxG) which are gene activators and Polycomb group (PcG) that are repressors (Ringrose and Paro, 2004). These two groups regulate gene expression in stem cell renewal, cell differentiation and cancer.

Opening of Chromatin Structure

SWI/SNF Complex

SWI/SNF complex is an important member of trxG group which opens chromatin structure and regulates transcription of genes. SWI derives from mutants with faulty mating-type switching in yeasts and SNF from yeast mutants, which did not ferment sucrose due to the defective transcription of the sucrose non-fermenting gene (*Suc2*) (Winston and Carlson, 1992). The complex, in addition to activation of the genes, may repress other genes (based on microarray studies). The genes regulated by this complex are spread throughout the genome (Sudarsanam et al., 2000; Roberts and Orkin, 2004). SWI/SNF contains 9 to12 subunits, but only 4 of them are required for in vitro activity. They include INI1 (SNF5), BRG1 (or BRM), BAF155 and BAF170 (Phelan et al., 1999). The composition of the complex also depends on the type of the cell. The complex hydrolyzes ATP and is using the energy of hydrolysis for chromatin remodeling.

The activities of SWI/SNF are conserved through eukaryotes (Roberts and Orkin, 2004), and are required for regulation of transcription of many target genes and also for DNA synthesis and regulation of mitosis (Flanagan and Peterson, 1999; Krebs et al., 2000). Loss of *INI1 function* due to mutations leads to highly malignant brain tumors including AT/RT (Rorke et al., 1995), choroid plexus carcinoma (Taylor et al., 2000; Gessi et al., 2003), meningioma (Schmitz et al., 2001) and primitive neuroectodermal tumors (PNET) (Sevenet et al., 1999). INI1 increases expression of *p16* and it might function upstream of RB in controlling malignant growth. RB mediated cell cycle arrest requires brahma-related gene 1

(*BRG1*) (Versteege, et al., 2002; Betz et al., 2002). SWI/SNF interacts with MYCC, RB, BRCA and MLL (mixed-lineage leukemia). RB forms complexes with SWI/SNF and HDAC in G1 phase of the cell cycle and represses transcription of cyclins E and A. Cyclin D – cyclin-dependent kinase 4 (CDK4) complex partially phosphorylates RB that allows release of HDAC and cyclin E which can now express its activity and cause G1-S transition. Higher concentration of cyclin E (transcribed by E2F) allows formation of the complexes with CDK2, further phosphorylation of RB and SWI/SNF and breakdown of the complex. This leads to activation of cyclin A and S-G2 transition (Trouche et al., 1997; Zhang et al., 2000).

Histone Replacement

Another possibility of activation of condensed chromatin is replacement of histone dimers H2A/H2B with H2A.Z/H2B. Variant protein H2A.Z prevents the spread of silenced chromatin. In yeast it prevents association of Sir2, Sir3 and Sir4 proteins with nucleosomes (Krogan et al., 2003; Mizuguchi et al., 2004). These complexes produce more accessible chromatin, not only through the exchange of proteins, but also through formation of nucleosomes without one histone dimer (Hirschhorn et al., 1992). A complex Swrl isolated from yeast and related to Swi2/Snf2 is playing a more important part (Mizuguchi et al., 2004). This complex consists of 12 proteins that associate with yeast variant (Htzl) of human histone H2A.Z (Krogan, et al., 2003; Mizuguchi et al., 2004). In the absence of H2A.Z, silencing spreads into the adjacent chromatin (Meneghini et al., 2003). It is possible that H2A.Z is modified through acetylation (Ren and Gorovsky, 2001). HAT complex NuA4, which is involved in the silencing of telomeric heterochromatin has four common subunits with Swrl (Boudreault et al., 2003) and human complex, which is equivalent to yeast Swrl, possesses also HAT activity (Owen-Hughes and Bruno, 2004) (see section *Replicative Clock).*

Closing of Chromatin Structure

PcG genes control repression of *HOX* in *Drosophila melanogaster* (Simon and Tamkun, 2002). Through their action they preserve the development of body patterning in *Drosophila,* but in mammals they control stem cell renewal, cell cycle and cancer (Lessard and Sauvageau, 2003). Chromatin compaction can be induced by the products of PcG group of genes, myeloid and erythroid nuclear termination stage – specific protein (MENT), Polycomb Repressive Complex 1 (PRC1), and complexes which bind N-terminal "tails" of the histones. MENT condenses chromatin by bridging the linker DNA and binding to H1 histones (Bednar et al., 1998). A similar mechanism was proposed for condensin and MeCP2 (Georgel et al., 2003, Springhetti et al., 2003; Porter et al., 2004). PRC1 is one of PcG complexes which inhibits transcription and chromatin remodeling (Shao et al., 1999; Francis et al., 2001). PRC1 core complexes (PCCs) are instrumental in producing highly compacted chromatin (Francis et al., 2004, Lavigne et al., 2004). Such action however requires nucleosomes but not histone tails (Francis et al., 2004) and each complex can compact three nucleosomes. PSC which is one of the subunits of PRC1 inhibits chromatin remodeling and transcription (King et al., 2002). Complexes of HP1 and repressors bind to protruding histone tails (Ducker and Simpson, 2000; Thiru, et al., 2004). PCC and HPI stabilize higher order folding, whereas SWI/SNF and HATs destabilizes closed chromatin 30 mm fiber.

Histone Modification

Acetylation

Histone modification is closely connected to chromatin remodeling (Jaenisch and Bird, 2003). Sequences in the histone tail can interact with DNA and other nucleosome proteins to mediate the coiling of the nucleosomes into more compact chromosomal structures. Lysine residues in the histone tails are selectively modified through acetylation and methylation, while phosphorylation modifies serine residues. Acetylation neutralizes positive charge of lysine and weakens the interaction of the tail with DNA and also prevents nucleosomes from folding into stable fibers. On the other hand, histone acetylation helps gene expression by facilitating the binding of RNA polymerase. Deacetylation of the histone prevents gene expression. Genes which are not active do not contain acetylated histones. For example, the entire inactive X chromosome of female mammals is almost completely deacetylated.

Other Modification Changes

Until recently, it was thought that the main factors in histone modification were differences in electrical charges between nucleosomes and DNA which allow to loosen or tighten the bond and influence silencing or activation of the genes. It is now known that histone acetylation plays a more important part than simply changing the electrical charge on the histone tails. In addition to acetylation, other modification changes are playing an important part, including methylation, phosphorylation, ubiquitination, connection with DNA methylation, RNAi and LTRs (Lachner et al., 2001; Bannister et al., 2001; Maher and Reik, 2001). Modification of histones creates codes that are different for silent and active genes and which tell the cells when to turn genes on and off (Rea et al., 2000; Strahl et al., 1999). Histone codes help preserve cellular identity of heritable patterns of gene expression which is known as cellular or transcriptional memory (Sims and Reinberg, 2004). Active euchromatin is characterized by histone hyperacetylation, methylation of lysine 4 on histone H3 (H3K4) and no methylation of CpG. Inactive heterochromatin has deacetylated histones, methylated H3K9 and methylated DNA. Methylated H3K9 is recognized by HP1 which also binds DNMT and histone methyltransferase (HMT). Subsequent methylation is facilitated by HDAC which deactylates lysine in preparation for methylation. DNMT and MBP methylate DNA and interact with HDAC and HMT. Additional DNA-binding proteins, such as zinc finger transcriptional repressors and corepressors, are recruited (Feinberg and Tycko, 2004; Zelent et al., 2005). Some HMT's function as suppressors of neoplastic growth (Huang, 2002). Recently isolated histone demethylase silences the genes (Shi et al., 2004).

Amino Acid Derivatives and Carboxylic Acids which Restore Normal Gene Expression

Derivatives of Glutamine and Phenyl-Spacer Compounds

The mechanisms which contribute to gene silencing in aging and cancer are reversible. They can be modulated by pharmaceutical agents, diet and dietary supplements and even cosmetics. There is hope that introduction of new agents which activate silenced genes and decrease the activity of abnormally over-expressed genes, may provide control of cancer and slow down the aging process. Simple amino acid derivatives and organic acids may help to activate silenced tumor suppressor genes *TP53, p21, PTEN* and *INI1* and decrease overexpression of oncogenes *RAS, AKT* and *MYCC* (Burzynski, 2004a). The derivatives of L-glutamine: 3-phenylacetylamino-2, 6- piperidinedione (A10), sodium salts of phenylacetylglutamine (PG) and phenylacetylisoglutamine (isoPG) and sodium salts of carboxylic acids, such as phenylacetate (PN) and phenylbutyrate (PB) help accomplish this task **(Figure 1)**. They are present in human blood and urine, dairy products and vegetables, and have been reproduced through organic synthesis (Burzynski, 2004a; 2004b; 2005; Burzynski et al., 2005).

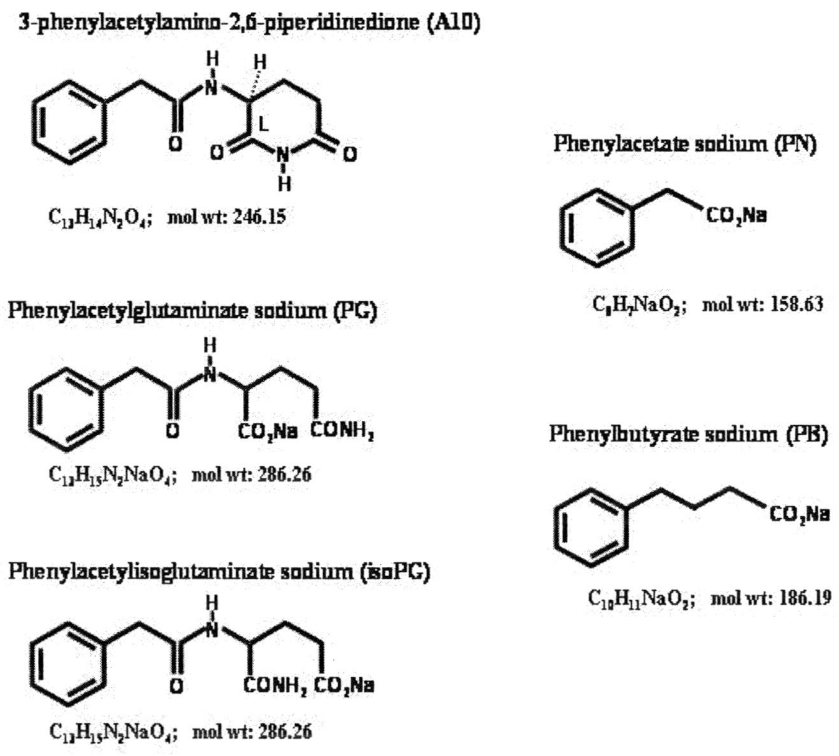

Figure 1. Chemical Formulas of Derivatives of Glutamine and Phenyl-Spacer Compounds.

3-phenylacetylamino-2, 6-piperidinedione

A10 protects the sequences which are vulnerable to the effects of carcinogens such as aflatoxin B_1 and benzo[a]pyrene through specific intercalation with DNA (Burzynski et al., 1983; 1984; Lehner et al., 1986; Hendry et al., 1987). Diet containing A10 protects mice and rats from development of breast, lung and liver cancers after exposure to benzo[a]pyrene, urethane and aflatoxin B1 (Kampalath et al., 1987; 1990; Muldoon et al., 1987; 1990; Hendry and Muldoon, 1988; Eriguchi et al., 1988). After ingestion, A10 is partially hydrolyzed in the small intestine into PG and isoPG (Ashraf et al., 1986; Burzynski et al., 1986).

A10 is available as an ingredient of a dietary supplement Aminocare® A10 (Burzynski, 2004b; 2005). The users of this supplement reported anecdotal positive effects which indicate possible application in age management regimen. From another point of view, A10 is a paternal compound for PG and isoPG and some of its pharmacological actions can be attributed to these two metabolites of A10 (see section Clinical Study in Skin Aging).

Phenylacetylglutamine and Phenylacetylisoglutamine

Human and sub-human primate livers have the ability to synthesize PG from L-glutamine and PN (James et al., 1972; Webster et al., 1976). Gene array studies revealed that PG effects expression of genes playing an important part in aging and cancer. (Waldbillig and Burzynski, 2003; Waldbillig et al., 2004).

Effect on *AKT* Oncogene

Gene *AKT* is a cellular homolog of retroviral *V-Akt* and encodes three different AKT enzymes (AKT1, AKT2, AKT3), which are serine-threonine kinases. Increased signaling through *AKT* pathway occurs frequently in high-grade malignancies (Vivanco and Sawyers, 2002; Zhu and Parada, 2002). AKT, which is also known as protein kinase B (PKB), regulates various cellular processes, such as proliferation, growth, apoptosis, and cytoskeletal rearrangement. AKT mediates cellular targets after activation through phosphorylation and blocks p53 protein and apoptosis of cancer cells through translocation of MDM2 into the nucleus (Mayo and Donner, 2001). AKT stabilizes cyclin D and induces cytoplasmic localization of p21 and KIP1 (or p27), which stimulates progression through the G1 phase of the cell cycle (Zhou et al., 2001; Weng et al., 2001). PG regulates AKT through a decrease of expression of *AKT2*.

Effect on *MYCC* Oncogene

The increase of expression of MADS box transcription enhancer factor (*MEFD2D*) by PG facilitates induction of MAD (Waldbillig and Burzynski 2003; Waldbillig, et al., 2004). MAD protein regulates the activity of MYCC-MAX complex and allows terminal differentiation of cancer cells. *MYCC* is over-expressed or amplified in numerous malignancies (Pelengaris et al., 2002; Herms et al., 2000). *MYCC* promotes cell growth and proliferation, inhibits terminal differentiation and sensitizes cells to apoptosis, if the proliferative pathway or antiapoptotic proteins are blocked. *MYCC* through MYCC-MAX dimers promote G1-S progression by activation genes *CCND2* (encoding cyclin D2) and *CDK4*. This leads to sequestration of p27 which is an inhibitor of CDK2. p27 undergoes ubiquitination promoted by *CUL-1* and *CKS* which are the genes targeted by MYCC. Under such conditions, p27 is no longer inhibiting cyclin E-CDK2 complex which permits phosphorylation of this complex by cyclin-activating kinase (CAK) and entry to G2. MYCC-MAX represses p15 and p21 (that prevent cell cycle progression) that are inhibitors of cyclin-E/CDK2 (Pelengaris et al., 2002). MYCC-MAX, in association with MYCC-interacting zinc finger protein 1 (MIZ1), prevents the transactivation of *CDKN1A*, which encodes p21 and *CDKN2B*, which encodes p15. Such growth promoting activity of MYCC is inhibited by PN which activates *CDKN1A* through demethylation of its promoter (Pelengaris et al., 2002). MYCC-MAX complex also inhibits differentiation through chromatin remodeling. This action is blocked by MAD protein, which replaces MYCC in MYCC-MAX complex and binds MYCC-MAX target genes. Induction of MAD is facilitated by MEFD2D. PG is increasing expression of MEFD2D in a dose-dependent manner and causes terminal differentiation. MYCC-induced apoptosis is an independent process from MYCC-promoted cell proliferation (Waldbillig et al., 2004; Pelengaris et al., 2002).

Effect on *INI1* tumor suppressor gene. An important tumor suppressor gene *INI1,* located at 22q11, undergoes point mutation in highly malignant brain tumors including AT/RT and glioblastoma multiforme. The mutation unmasked nuclear export signal (NES) and causes INI1 protein to dislocate to the cytoplasm (Robert and Orkin, 2004). In normal cells, NES masking occurs by interaction with the proteins own carboxy-terminal NES inhibitory region. In some highly malignant brain tumors, for instance AT/RT, the mutation consists of premature stop condon in *INI1* that deletes the NES inhibitory region (Robert and Orkin, 2004). INI1 represses cyclin D1 transcription and induces p16, inhibiting G1-S cell cycle transition (Zhou et al., 2001). Reduced activity of INI1 leads to global hypomethylation and interferes with chromatin remodeling (described in section *Hypomethylation and Opening of Chromatin Structure*) Reconstitution with wild type *INI1*, the *INI1* deficient cells causes restoration of G1 cell-cycle arrest, apoptosis and repression of cyclin D1 and E2F. Cancellation of INI1 transport from nucleus to cytoplasm may depend on the RANGTP/RANGDP gradient and Ran binding protein 1 (RANBP1), which increases the amount of RANGDP in the cytoplasm. Based on DNA microarray studies by our group, PG decreases expression of *RANBP1* in a dose-dependent manner (Waldbillig and Burzynski, 2003; Waldbillig et al., 2004); therefore, PG may change the RANGTP/RANGDP gradient to prevent escape of INI1 from the nucleus.

Effect on Transforming Growth Factor β Gene. TGFB is involved in a complex network of several hundred genes which could be activated or suppressed. The response to TGFB depends on the type of cell and cell micro-environment; it can suppress neoplastic growth, but it can also cause tumor progression (Siegel and Massague, 2003). Increased biosynthesis of TGFB1 occurs in high-grade tumors. It promotes invasive and metastatic properties of neoplastic cells and creates an environment necessary for angiogenesis (Siegel and Massague, 2003). It also down-regulates expression of *PTEN* (Li and Sun, 1997). PG decreases expression of *TGFB1* in a dose-related manner (Waldbillig and Burzynski, 2003; Waldbillig et al., 2004).

Effects of Phenylacetylisoglutamine. The studies on pharmacology of isoPG are not as extensive as they are on PG, but the effects of isoPG in cell cultures are similar to the activity of PG. The effect of isoPG on expression of genes is currently being studied.

Phenylacetate

PN occurs in dairy products and human blood, and in trace amounts in vegetables (Burzynski et al., 2005). It affects the activity of *RAS* and *BCL2* oncogenes and *TP53, p21* and neurofibromatosis 1 *(NF1)* tumor suppressor genes.

Effect on *RAS* Oncogene

The *RAS* oncogene family plays an important part in aging and is over-expressed in more than 30% of all cancers. It forms, perhaps, the most important signal transmission pathway in cancer (Malumbres and Barbacid, 2003). Knowledge about RAS signaling has expanded impressively over the last 30 years. However, a complete understanding of RAS function is still to come. RAS proteins are activated by various extracellular stimuli, which are mediated by receptor tyrosine kinases (RTK), G proteins, adhesion molecules and second messengers. Active RAS proteins induce a wide variety of cell responses, such as transcription, translation, cell cycle progression and apoptosis or cell survival (Malumbres and Barbacid, 2003). PN interrupts signal transduction in the RAS pathway through the inhibition of farnesylation of the RAS protein (Castillo et al., 1991; Adam et al., 1995). The detailed mechanism involves inhibition of mevalonate 5-pyrophosphate decarboxylase (Shack et al., 1995; Samid et al., 1992).

Effect on *NF1* Tumor Suppressor Gene

NF1 is not the subject of extensive research in aging, but mutations of *NF1* an its loss of functions have been studied for many years in the area of brain tumors and neurofibromatosis. In high-grade gliomas there is substantial evidence of mutations of *RAS*, but such mutations have not been identified in low-grade astrocytomas (Zhu and Parada, 2002). In these tumors, however, the inactivation of RAS signaling can be accomplished through neurofibromine, which is a product of *NF1* (Guha et al., 1997). Neurofibromine hydrolyzes GPT-RAS and inactivates it. Mutation or silencing of *NF1* leads to activation of RAS.

Effect on *BCL2* Oncogene

Efficient programmed destruction and removal of abnormal and malfunctioning cells through apoptosis is central to protection against cancer and important in age management. PN down-regulates anti-apoptotic BCL2 and PG may also inactivate anti-apoptotic protein BCL-X_L. This leads to the release of cytochrome C from mitochondria and triggers apoptosis (Adam et al., 1997).

Effect on *TP53* Tumor Suppressor Gene

The tumor suppressor gene *TP53* plays a crucial part in the regulation of cell cycle in normal, aging and cancer cells. p53 protein works as a transcription factor, which activates numerous genes (Vousden and Lu, 2002). Theoretically, the human genome contains over 4,000 p53 binding sites. There are eighteen possible posttranslational modifications of p53 including phosphorylation and acetylation. The genes transactivated by p53 regulate inhibition of cell growth, DNA repair, activation of apoptosis and regulation of angiogenesis. Among crucial functions of p53 is induction of apoptosis, which starts with mitochondrial release of cytochrome C (Vousden and Lu, 2002). The action of p53 is blocked by AKT through MDM2, but p53 activates the transcription of *PTEN* and down-regulates *AKT*. It is estimated that loss of *TP53* function through mutations or silencing occurs in over 50% of all malignant tumors (Vousden and Lu, 2002). PN activates *TP53* through demethylation of its promoter (Liu and Samid, 1995), but PG activates p53 through the decrease of expression of *AKT2*. Gene *CDKN1A* which encodes protein p21 is a direct transcriptional target of p53. DNA damage activates TP53, which transactivates *CDKN1A*. This leads to release of p21 which inhibits the complexes cyclin D/CDK4 and cyclin E/CDK2 resulting in cell cycle arrest at G1/S and apoptosis. p21 inhibits DNMT and may activate a number of tumor suppressor genes, which are silenced through hypermethylation of their promoters (Gorospe et al., 1996).

Phenylbutyrate

PB occurs in dairy products and vegetables, and is partially converted in the liver to PG and PN. In addition to effects typical for PG and PN, it activates tumor suppressor genes through inhibition of histone deacetylase (Collins et al., 1995; Kamitani et al., 2002).

Multiple Targets and Mechanisms of Action

A10, PG, isoPG, PN and PB effect multiple targets and have different mechanisms of action. Research data indicate that PG activates tumor suppressor genes *MAD* and *PTEN* and interferes with signaling in *AKT2* and *MYCC* pathways. It blocks expression of *TGFB1* and normalizes nuclear transport through decrease of the expression of *RANBP1* which may restore the activity of mutated INI1 protein **(Figure 2)**. PN interferes with signal transmission in RAS and BCL2 pathways and activates expression of tumor suppressor genes *TP53, p21* and *NF1*.

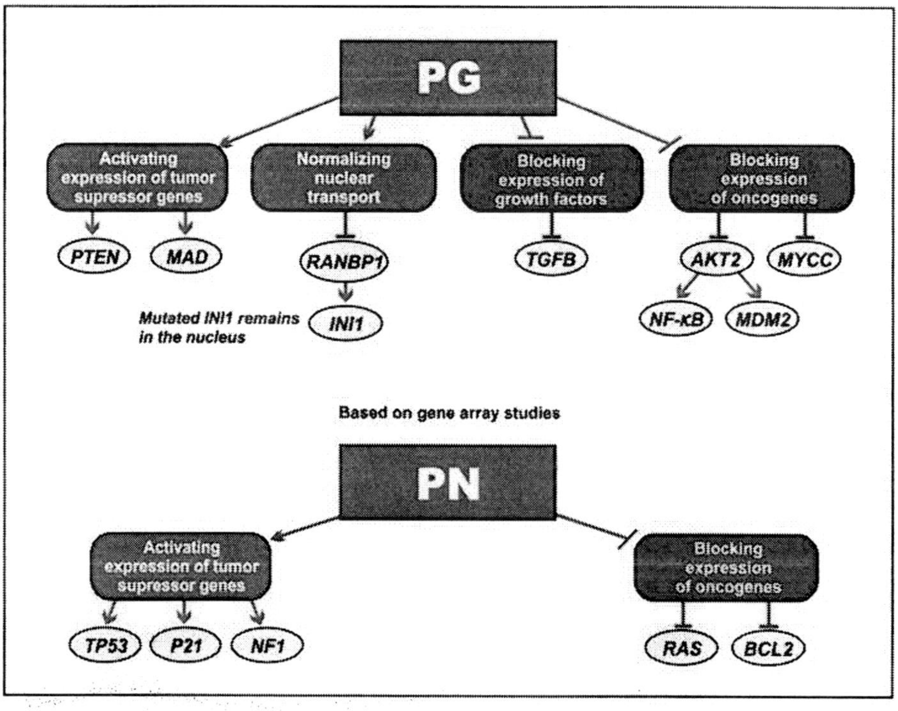

Figure 2. Multiple Targets of PG and PN. Based on research data, it is proposed that PG activates tumor suppressor genes *PTEN* and *MAD* and blocks expression of *TGFB1, MYCC* and *AKT2* (and indirectly *NF-κB* and *MDM2*). PG normalizes nuclear transport through decrease of expression of *RANBP1* which may restore the activity of mutated INI protein. PN activates expression of tumor suppressor genes *TP53, p21* and *NF1* interferes with signal transmission in *RAS* and *BCL2* pathways.

The mechanisms of action of PG and PN involve methylation of DNA and acetylation of histones **(Figure 3).** PN inhibits aberrant methylation of promoters regions of tumor suppressor genes and deacetylation of histones which leads to activation of silenced tumor suppressors. PG normalizes global hypomethylation of the genes which decreases DNA instability and normalizes gene expression through RNAi mechanism. It also affects chromatin remodeling, possibly through INI1.

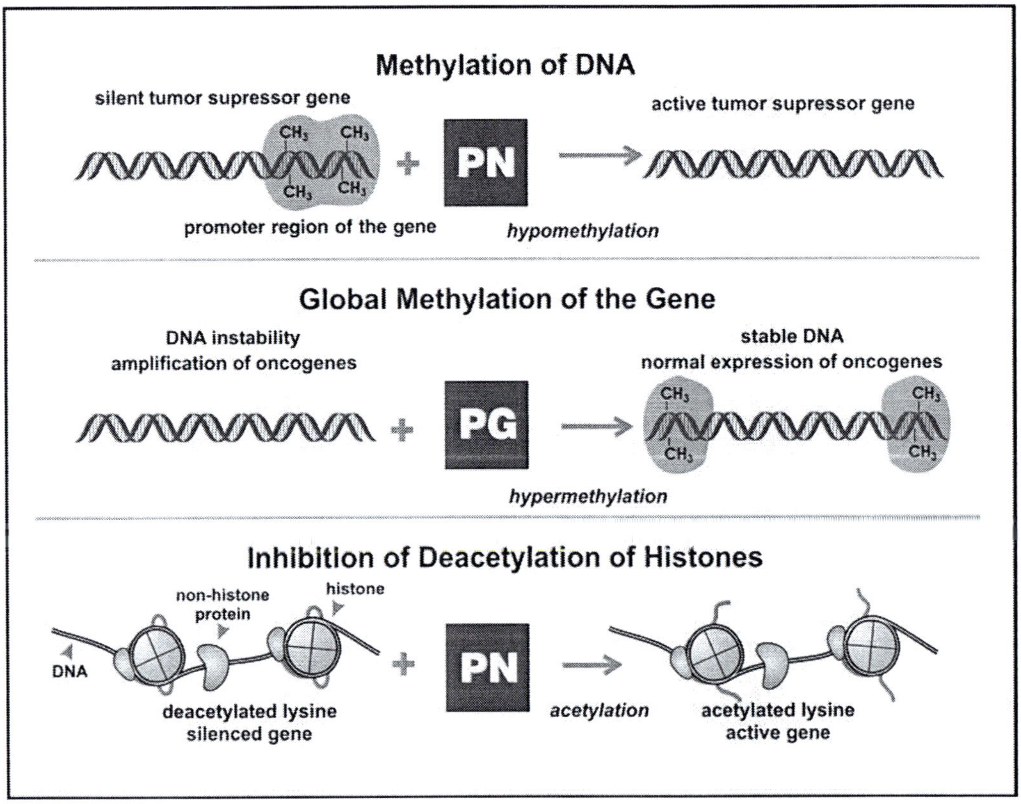

Figure 3. Different Mechanisms of Action of PN and PG. The mechanism of action of PN and PG involve methylation of DNA and acetylation of histones. PN inhibits aberrant methylation in promoters regions of tumor suppressor genes and deacetylation of histones. This leads to activation of silenced tumor suppressors. PG normalizes global hypomethylation of the genes which decreases DNA instability and normalizes gene expression through RNAi mechanism. It also affects chromatin remodeling possibly through INI1.

A10 and PB, which are metabolized to PG, isoPG and PN, share the most mechanisms of action with PG and PN.

Pharmaceutical and Nonprescription Formulations

PB is available in the United States and the European Union as a prescription drug and A10, PG and isoPG are ingredients of supplements and cosmetics. Together with selected amino acids and vitamin B2, A10 is an ingredient of Aminocare® A10 and PG of Aminocare® Extra (Burzynski, 2004b; 2005). PG and isoPG are active ingredients of cosmetic products, Aminocare® cream and Aminocare® lotion (Burzynski, et al., 2004).

Gene Silencing in Aging

New Theory of Aging

Gene silencing theory of aging propose that the key process in aging involves reduced expression of genes (Burzynski, 2003; 2004b; 2005; Burzynski et al., 2004). Based on research data in animals and humans coming from different laboratories, it is estimated that approximately from 0.5 to 4.5% of genes, active in young age, are silenced in the old organisms (Cao et al., 2001; Kyng et al., 2003; Geigl et al., 2004). Epigenetic mechanisms involving methylation of promoters regions and deacetylation of histones could silence the genes (Feinberg and Tycko, 2004; Ordway and Curran, 2002). Genome-wide hypomethylation (outside promoters) induces genomic instability and contributes to silencing of tumor suppressor genes and amplification of oncogenes. Environmental factors, dietary errors, habits, stress, pathological changes and genomic instability leads to profound changes in gene expression. At the age of 25, the body has the optimal combination of active genes, but soon thereafter various groups of genes are turned off as the result of aging leading to hormonal changes during menopause, graying and loss of hair, reduced immunity and detoxification and formation of cancer. Silencing of tumor suppressor genes triggers increased signaling through oncogene pathways, and increases the risk of cancer. Aberrant methylation of CpG islands in promoters regions is instrumental in silencing of the genes, but global demethylation contributes to amplification of oncogenes and silencing of tumor suppressors (Feinberg and Tycko, 2004; Ordway and Curran, 2002).

Research Which Supports Gene Silencing Theory

Changes of gene expression in aging were analyzed in different organisms including yeast, nematodes, insects, mice and humans (Cao et al., 2001; Kenyon, 2001; Howitz et al., 2003; Longo and Finch, 2003; Murphy et al., 2003; Kyng et al., 2003; Geigl et al., 2004). Different types of cells were studied such as lymphocytes, fibroblasts and hepatocytes. The analysis of 11,000 genes in aging livers of mice in comparison to young animals revealed that approximately 1% of the genes changed expression during aging and most of them were silenced (Cao et al., 2001). Tumor suppressor genes, *Pten* and *Igfbp1,* were silenced which increased signal transduction through *Igf* pathway. Other important groups included detoxification (*Gsto1*) and cholesterol metabolism genes (*Apoe*). Among a smaller group of genes, the most pertinent which have increased expression during aging were those associated with chronic inflammation (*C1qb*), stress response (*Hsp genes)* and protection from apoptosis (*Igf* through *Igfbp1* and *Api6).* According to other studies in different organisms, including yeast, nematodes, insects, and humans, in aging there is increased signaling through pathways corresponding to human *IGF1/AKT* pathway and decreased activity of tumor suppressors *PTEN* and *p53* (Kenyon, 2001; Howitz et al., 2003; Longo and Finch, 2003; Murphy et al., 2003; Kyng et al., 2003; Geigl et al., 2004). Important human studies were performed on fibroblasts from young and old donors and a premature aging disorder, Werner's Syndrome (WS) (Kyng et al., 2003). The analysis of 6,912 genes revealed that 6.3% had altered expression in cells taken from aging individuals and those diagnosed with WS compared with young donors. Changes in gene expression in normal aging were almost identical with WS. Most of the genes (70%) were down regulated, including tumor suppressor genes, inhibitors of apoptosis, transcription factors, genes involved in chromatin remodeling and RNA pol II complex, and there was increased expression of oncogenes (Kyng et al., 2003). In a different study, the analysis of expression of 14,500 genes in human fibroblasts and lymphocytes revealed differences in expression in 2.8% genes (Geigl et al., 2004). It is interesting to notice the increased activity of the genes playing an important part in silencing, such as *H2AFY2* involved in establishing silent chromatin domains, *MeCP2* coding MBP and *PER2* involved in circadian clock (Geigl et al., 2004).

Changes in Gene Expression in Skin Aging

Aging of Epidermis and Collagen

Background

Microarray studies on changes in gene expression in aging have focused on fibroblasts, lymphocytes and hepatocytes with less attention to cells of epidermis (Cao et al., 2001, Kyng et al., 2003; Geigl et al., 2004). Skin is the largest organ of the body and very important in aging. Changes during aging of the skin are easily visible and are of great concern to everybody, but they are also easily accessible for clinical observations and are considered the most obvious signs of aging (Giampapa, 2003). The skin is continuously exposed to assault from the environment including mechanical factors causing damage and wounds, ultraviolet radiation (UV) and numerous chemicals, which create reactive oxygen species (ROS) and promote photoaging. On the other hand, the skin is under the influence of endogenous factors which accelerate chronological (normal) aging of the entire body including a high calorie diet, lack of exercise, psychic stress, changes in circadian rhythm, hormonal and chronic pathological processes (such as inflammation and hyperglycemia). These adverse processes affect various parts of the skin, but most of all the epidermis and collagen, and cause most common signs of skin aging - wrinkles **(Figure 4)**.

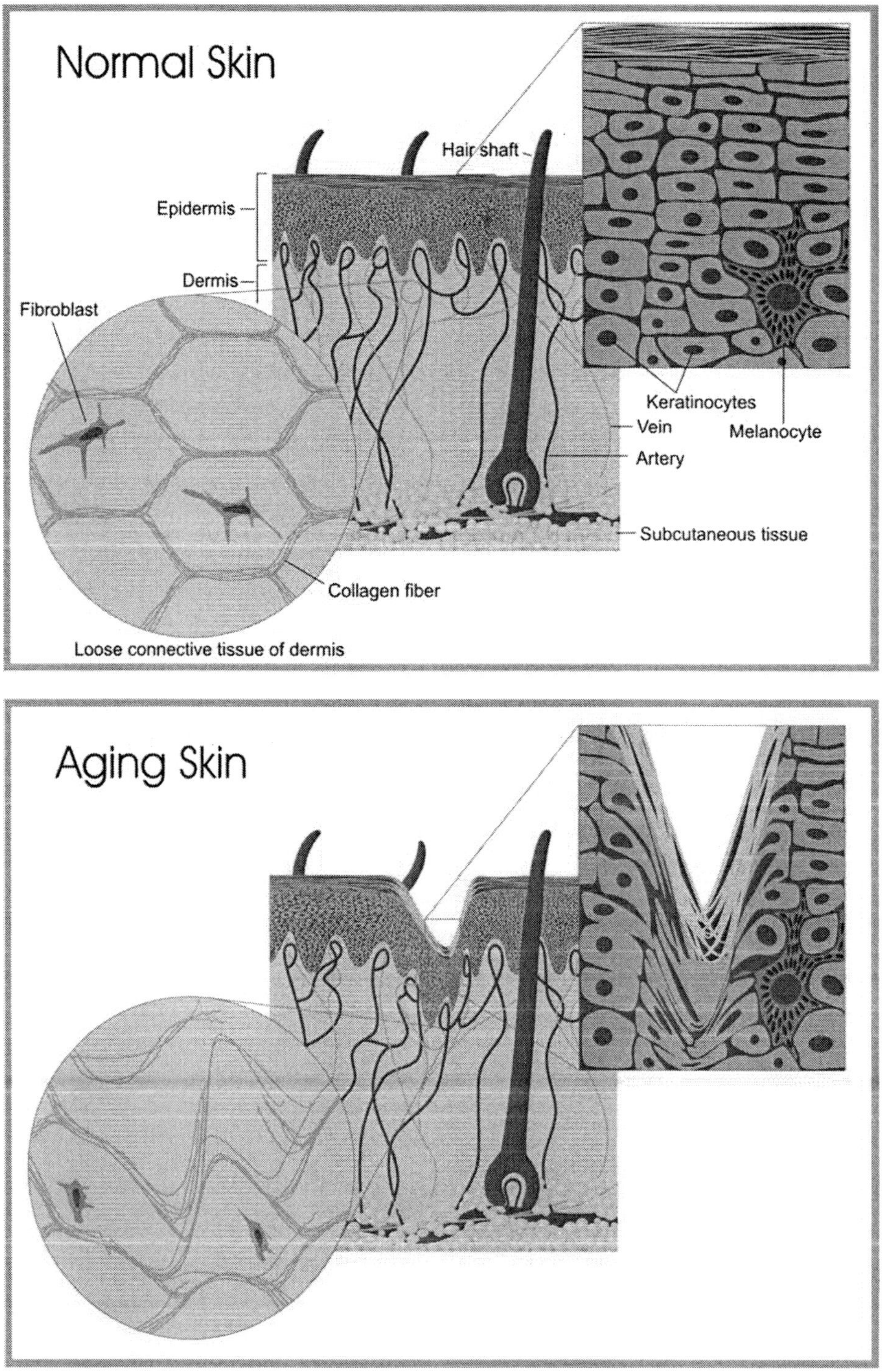

Figure 4. Aging of the Skin. During aging process, the epidermis is becoming thinner, dehydrated and contains an increase proportion of damaged cells. Changes in the epidermis are accompanied by destruction of skin collagen. Healthy collagen fibers create a fine mesh, which provides elasticity and softness to the skin. Disturbances in cell renewal and collagen repair leads to formation of wrinkles.

Renewal Process

To protect itself, the body uses a continuous process, which attempts to replace damaged and old cells. The turnover rate of epidermal cells is approximately 60 days in humans and 7 days in mice (Ghazizadeh and Taichman, 1998; Hunter et al., 1995). Renewal of the epidermis is accompanied by collagen formation by fibroblasts. Disturbance of the delicate balance between cell renewal, cell loss and replacement of damaged collagen fibers leads to acceleration of aging, precancerous changes and cancer. The renewal of the tissues depends on stem cells. Such cells are undifferentiated and are able to give origin to one or several lineages of tissue. These cells have clonogenic potential and are capable of self-renewal. They undergo asymmetrical cell division in which one cell differentiates down to specific lineage and the other cell remains a stem cell (Perez-Losada and Balmain, 2003).

Imbalance in Cell Renewal

Disturbances in cell renewal may accelerate aging and lead to numerous precancerous and cancerous changes (Perez-Losada and Balmain, 2003; Klein-Szanto et al., 1993). The imbalance in cell renewal is reflected by cell loss, which is not counteracted by cell division. The skin accumulates damaged and senescent cells which are not replaced by new cells and these cells are not able to efficiently absorb nutrients and expel waste materials. Damage to the cell membranes causes them to exclude water. The epidermis is thinner and dehydrated and contains an increased proportion of damaged cells contributing to wrinkles.

The changes in the epidermis are accompanied by destruction of skin collagen. Healthy collagen fibers create a fine mesh, which provides elasticity and softness to the skin. It is suggested that, in chronological skin aging, increased synthesis of matrix metalloproteinases (MMP) is instrumental in collagen destruction. This was also found in photoaging after UV radiation where increased expression of MMP genes is a major event causing connective tissue damage (Talwar et al., 2001; Chung et al., 2001). However, in photoaging, UV radiation stimulates collagen synthesis at the same time (Chung et al., 2001). This new collagen synthesis is not sufficient since the newly synthesized procollagen secreted from the fibroblasts is degraded by MMPs 1 and 2 (Fisher et al, 1997). In chronological aging, there is decreased formation of new collagen due to poor renewal of fibroblasts and increased fibroblast senescence, which is well illustrated in progeroid syndromes such as WS (Kipling et al, 2004). In WS, a substantial reduction of divisional capacity of skin fibroblasts to less than 20 population doublings versus normal of 40 to 100 indicates the importance of replicative senescence in aging (Salk et al, 1981). Damaged collagen fibers crosslink and become hard, lose water and collapse leading to formation of deep wrinkles.

ROS-induced damage to cellular membranes and chronic subclinical inflammation are major factors, which trigger profound changes in gene expression leading to skin aging (Perricone and Giampapa, 2003). In both chronological aging and photoaging, a chronic, subclinical inflammatory process is responsible for producing cytokines such as TGFB, which inhibits formation of the dermal extra-cellular matrix (Werth et al., 1997). Neutrophils activated during chronic inflammation produced by UV light, produce ROS which could damage adjacent keratinocytes (Trush and Kensler, 1991). Such damage effects membranes and causes DNA-strand breaks, DNA adducts and alterations in gene expression. Changes in expression of oncogenes play an important part in the renewal of the epidermis and induction of neoplastic growth.

Premature Senescence Caused by Oncogenes

P13K Pathway

UV light and ROS mediate activation of two important pathways: PI3K and p38 mitogen-activated protein kinase (p38 MAPK – p38). (Figure 5)

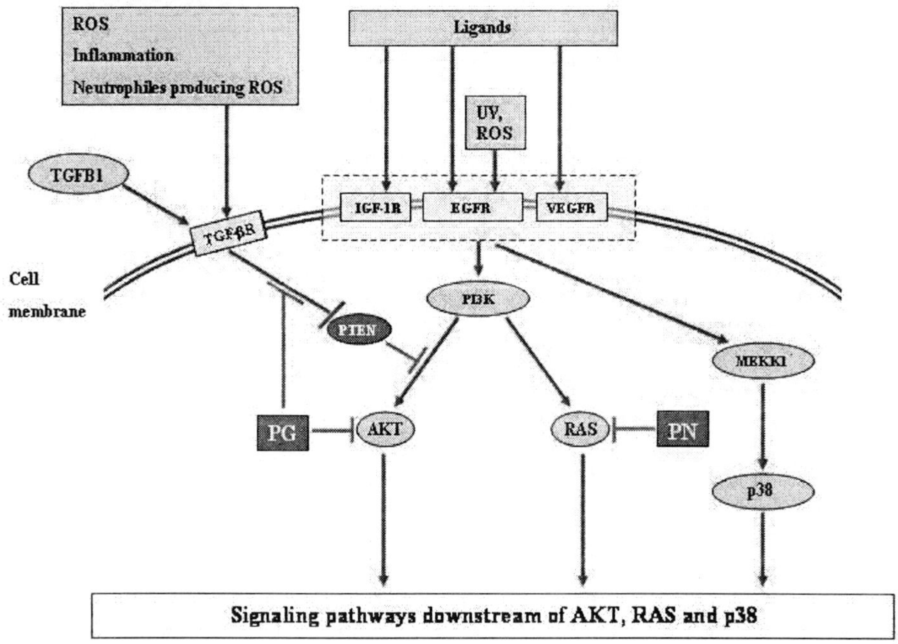

Figure 5. Premature Senescence Caused by Oncogenes (I). ROS, inflammation, neutrophils producing ROS, UV light and ligands binding to growth factor receptors activate TGFB1, PI3K, AKT, RAS and p38 pathways. PG interrupts signal transduction in AKT pathway through decrease of expression of *AKT2* and through activation of PTEN. PN interrupts signal transduction in RAS pathway through inhibition of farnesylation of RAS protein.

Numerous animal studies on aging have revealed the crucial role of PI3K (Burzynski, 2005a). PI3K is one of the most important components in the IGF/AKT pathway, which was proven by numerous studies to play a crucial part in aging (Burzynski, 2004b; 2005; Burzynski et al., 2004). Activation of PI3K triggers a complex cascade of events. One signaling pathway leads through AKT, mTOR, glycogen synthase kinase – 3 β (GSK-3β) and cyclic-AMP-response-element-binding protein (CREB) (Bowden, 2004). (Figure 6)

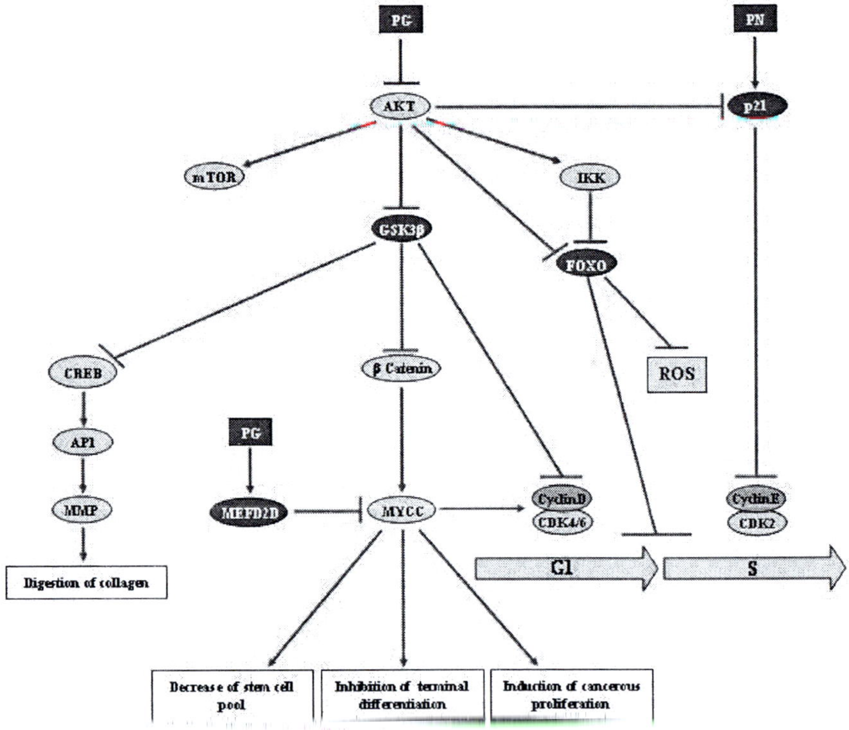

Figure 6. Premature Senescence Caused by Oncogenes (II). AKT activates mTOR and IKK which may promote cancer and aging and inhibits FOXO, p21 and GSK3β which protect against cancer and slow down aging process. Inhibition of GSK3β by AKT results of activation of CREB, AP1 and MMP which leads to digestion of collagen and formation of wrinkles. Activation of MYCC may result in decrease of stem cell pool, inhibition of terminal differentiation and induction of cancerous proliferation. Inhibition of p21 and FOXO by AKT induces G1/S transition. Detoxification of ROS is impaired through inhibition of FOXO by AKT and IKK. PG controls these negative changes through decrease of expression of *AKT2* and inhibition of MYCC and PN through activation of p21.

Another important aspect of the effect of PI3K/AKT pathway in aging is inactivation of FOXO factors. FOXO factors, FOXO1, FOXO3A and FOXO4, are localized in nucleus and activate expression of the genes which prevent proliferation (Burzynski, 2005). Activated AKT phosphorylates FOXO factors and changes their localization from the nucleus to cytoplasm, which stimulates cell proliferation (Vivanco and Sawyers, 2002). (Figure 6) In addition to PI3K/AKT, another pathway is involved in phosphorylation of FOXO3A and its exclusion from the nucleus – I kappa B kinase β (IKK)/NF-κB. According to Hu et al., IKK phosphorylates FOXO3A independently of AKT (Hu et al., 2004). This leads to ubiquitin-dependent proteolysis of FOXO3A.

RAS Pathway

The other pathway activates RAS protein. UV irradiation also activates RAS through epidermal growth factor receptor (EGFR) without ligand involvement (Bowden, 2004). Mutant RAS induces papillomas and squamous cell carcinomas (Brown et al., 1998).

Phosphorylation of CREB allows its binding to the c-FOS promoter through c-FOS AP-1 (FAP1) and CRE sites and to the cyclooxygenase 2 (COX2) promoter (also through the CRE site).

Such binding to CRE and FAP1 sites enhances transcription of c-FOS and COX2 genes. Subsequently, c-FOS forms a complex with JUND (AP-1 complex) which binds to specific DNA elements (Bowden, 2004). AP-1 complex induces expression of the genes for collagen-digesting enzymes, such as MPP (Chung et al, 2001). c-FOS is expressed at low levels in many different types of cells (Saez et al, 1995). A number of different stimuli may induce c-FOS through CRE, FAP1 and serum response element (SRE) but CRE and SRE are the two elements in the promoter region of c-FOS, which play the most important part in induction of c-FOS in photoaging (Buscher et al., 1988). (Figure 7)

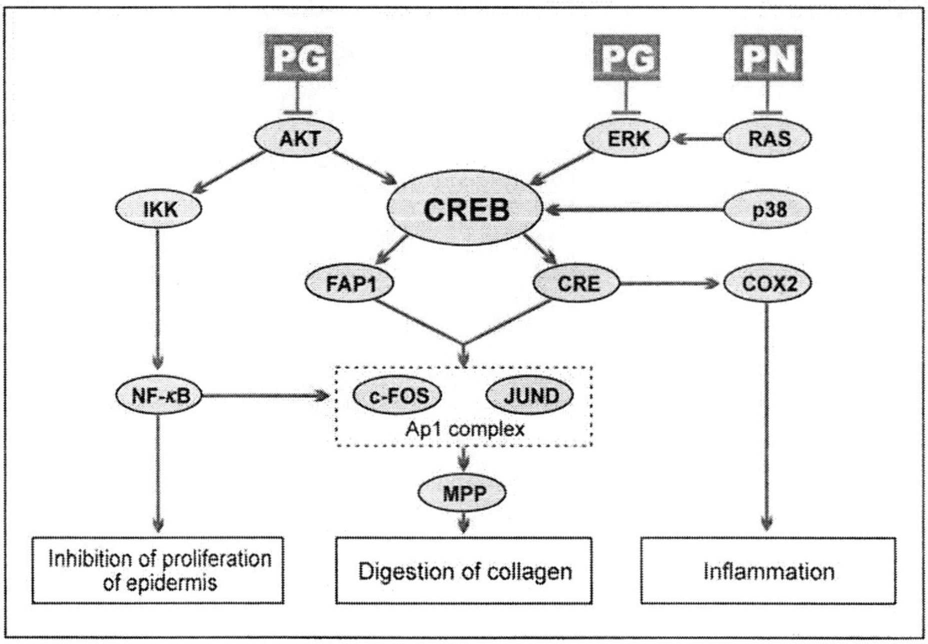

Figure 7. Premature Senescence Caused by Oncogenes (III). AKT leads to inhibition of proliferation of epidermis through phosphorylation of IKK (which releases active NF- κB), and digestion of collagen through activation of CREB, AP1 and MPP. RAS signaling pathway leads to digestion of collagen through ERK, CREB, AP1 and MPP and to increase of inflammation through ERK, CREB, CRE and COX2. PG decreases expression of *AKT2* and PN interrupts signal transduction in RAS pathway.

p38 Pathway and Photoaging

Another pathway more typically affected by UV light leads to p38 and extracellular-signal-regulated kinase (ERK). In natural aging, ERK is activated through the RAS pathway as mentioned before. p38 and ERK produce activation of CREB, which transactivates c-FOS.

The p38 pathway may play a more important part in c-FOS expression in human keratinocytes in photoaging (Bowden, 2004). (Figure 7)

In the PI3K pathway, the most pronounced effect on c-FOS transcription has activation of AKT and inactivation of GSK3β (Bowden, 2004). (Figure 6) In photoaging, in the p38 as well as in PI3K pathways, the initiation of UV light signaling occurs through activation of the EGFR which further transduce the signal through P13K and p38. (Figure 5) This activation does not, however, take place as usual through the ligand-binding mechanism, but through generation of ROS (Wan et al., 2001; Zhang et al., 2001), which activate p38 and PI3K.

NF-κB Pathway

NF-κB transcription factors form dimers (RelA/p65:p50), which in unstimulated cells, are bound to inhibitor kappa B (IκB) and remain in cytoplasm (Ghosh et al., 1998). A number of external stimuli, which include AKT, activate IKK signalsome (Li and Verma, 2002; Ducut et al., 2004). IκB and NF- κB undergo phosphorylation (DiDonato et al., 1997; Nelson et al., 2004). After phosphorylation, IκB is degraded by the proteasome and free NF-κB is translocated to nucleus, where it regulates transcription of different genes involved in aging, carcinogenesis, apoptosis, development and immune response (Li and Verma, 2002). In skin aging NF- κB inhibits proliferation of the epidermis (Seitz et al., 1988). Inflammatory compounds, which are created in this process, produce more ROS, which oxidize membranes surrounding the cells, mitochondria and nucleus. IκB is also among genes activated by NF-κB (Sun et al., 1993). (Figure 5) This creates a negative feedback loop, which forms a biological clock (Young and Kay, 2001; Nelson et al., 2004). NF-κB oscillates between cytoplasm and nucleus and assures a concentration of activated transcription factor in the nucleus necessary for proper response (Nelson et al., 2004). In aging organisms, this biological clock runs faster. Among other important oncogenes, the MYCC depletes the stem cell pool, and in the committed population of cells, inhibits terminal differentiation and induces proliferation (Waikel et al., 2001). (Figure 6)

The Effect of Amino Acid Derivatives and Carboxylic Acids

Based on gene array studies, PG decreases expression of AKT2, TGFB1 and increases expression of MEFD2D (Waldbillig and Burzynski, 2003; Waldbillig et al., 2004). The increase of expression of MEFD2D may indirectly decrease activity of MYCC (through MAD). AKT, TGFB1 and MYCC form important nodes of oncogene networks involved in aging and cancer. Through decrease of expression of TGFB1, PG may activate PTEN which is an inhibitor of AKT and that creates double interruption of signaling through IGF/AKT pathway (through PTEN and directly through AKT), which was found to be very important in aging. Interruption of signal transduction in this pathway was instrumental in substantial life extension in various organisms. Based on the research of various groups, this pathway was interrupted during caloric restriction (Kenyon, 2001; Cao et al., 2001). Among benefits which may help in skin aging are stimulation of stem cell pool, terminal differentiation and prevention of damage to collagen, inflammation and neoplastic transformation. (Figure 5-7)

Biological Clocks in Skin Aging

Circadian Clock

Background

A common feature of living organisms are rhythmic oscillations known as the circadian clock (Fu and Lee, 2003). Less than 24-hour oscillations, named ultradian, are observed in fast growing (cancerous) cells (Klevecz et al., 1987; Klevecz and Braly, 1991). Different endogenous clocks (division clocks) are controlling cell division in addition to circadian pacemakers. These biological clocks are based on molecular mechanisms which organize the behavior and physiology of the organisms around cyclic environmental changes, regulate gene expression and numerous other functions.

The circadian clock is centrally regulated through suprachiasmatic nuclei (SCN), which are located in the anterior part of the hypothalamus (Stephan and Zucker, 1972). This advanced system contains input and output pathways and the central pacemaker (Fu and Lee, 2003). Information from the environment is transmitted through the input pathways to the central pacemaker where it is processed to endogenous oscillations. The output pathways carry these oscillations to different targets controlling important processes (Reppert and Weaver, 2002). The rhythms generated in SCN are transmitted to peripheral tissues where they control cell proliferation, apoptosis and aging. Control is exerted through sympathetic and parasympathetic nervous systems, pituitary, adrenal and gonadal hormones, and other systems.

Genetic Control of Circadian Clock

A group of eight genes controls the circadian clock in SCN and in peripheral tissues: *Period 1 (PER 1)* identified in 1980s, *Period 2 (PER 2), Period 3 (PER 3), Bmal 1, Clock, Casein Kinase Iε (CKIε), Cryptochrome 1 (CRY 1)* and *Cryptochrome 2 (CRY 2)* (Tei et al., 1997; Sun et al., 1997; Shearman et al., 1997; Takumi et al.,1998; Hogenesch et al., 1998; King et al., 1997; Takano et al., 2000; Hsu et al., 1996).

The SCN neurons maintain a steady concentration of CLOCK protein. BMAL1 concentrations are the highest at the beginning of the day and lowest at the end of the day. BMAL1, when present in high concentration, heterodimerizes with CLOCK. The heterodimers BMAL1-CLOCK inhibit transcription of BMAL1. In the beginning of the day, they activate transcription of CRY and PER through binding to E box (CACGTG) of *CRY* and *PER* and at the end of the day, an increased amount of CRY1 in cytoplasm forms complexes with CKIε and PER proteins. In the beginning of the night, CKIε – PER-CRY 1 complex migrates to nucleus and disintegrates CLOCK/BMAL1 complex, inhibits transcription of CRY and PER and activates transcription of BMAL1 (Preitner et al., 2002; Etchegaray et al., 2003). *PER* genes encode PAS domain proteins which do not bind directly to DNA, but create a platform for formation of heterodimers of other clock proteins. CLOCK and BMAL1 represent a positive portion of the feedback loop and induce expression of *PER* and *CRY* (Gekakis et al.,1998). Proteins PER and CRY are components of the negative portion of the loop through E box enhancer elements targeted by CLOCK/BMAL1 (Kume et al., 1999). The activity of numerous genes is regulated by circadian clock and the activity varies depending on the organ studied. In mice, they include over 100 genes in the SCN and 400 in the liver, including receptors and their ligands, transcription factors and enzymes (Ueda et al., 2002).

Central Clock vs. Peripheral Clocks

The central clock controls peripheral clocks, but only to a certain point. Peripheral clocks can operate independently of the circadian clock to react quickly in order to exercise cell cycle control, repair DNA damage, inhibit cancerous growth or induce apoptosis (Fu et al., 2002). In conditions of caloric restriction, the peripheral clock may become independent from SCN. This is regulated through the supply of nutrients (Damiola et al., 2002; Stokkan et al. 2001) and in this respect the peripheral clock is influenced by cellular metabolism (Rutter et al., 2002).

Regulation of Signaling Pathways by Circadian Clock

The signal from the central clock regulates cell cycle and apoptosis in peripheral tissues through release of growth factors, corticosteroids, peptide hormones and neurotransmitters rather than responding directly to light stimuli (Ayroldi et al. 2002; Herrlich, 2001; Williams et al., 2001; Lavoie et al., 2002). **(Figure 8)**

Growth factors which are controlled by circadian rhythm activate RAS/ERK and MAPK1/2 pathways (Johnson and Lapodat, 2002). MAPK1/2 activate CREB which is responsible for activation of the transcription of core circadian genes in the tissues (Balsalobre et al., 2000). Peptide hormones and neurotransmitters control signaling through protein kinase A (PKA) which activates AKT inhibiting GSK3β and preventing inactivation of β catenin by GSK3β. β catenin promotes cancer growth through MYCC and cyclin D1. GSK3β is a homolog of core circadian gene "Shaggy" in Drosophila (Martinek et al., 2001). *MYCC* expression shows circadian rhythm in mouse tissues and this is regulated by PER2. Mutations of *PER2* caused increase of expression of MYCC and formation of lymphoma in mice after gamma radiation (Fu and Lee, 2002). In irradiated *PER2* mutant thymocytes, there is increased activation of *TP53* (Fu and Lee, 2002). Similar mechanisms involving clock genes occur in human keratinocytes after exposure to UV light (Kawara et al., 2002). Genes *MYCC* and *TP53* are connected with *PER* through the mechanism involving casein kinase II (CKII). CKIIα phosphorylates PER in *Drosophila,* which allows its nuclear transport, and phosphorylates and activates MYCC. p53 is also phosphorylated by CKII after exposure to UV light and decreases DNA damage response (Lin et al., 2002; Channavajala and Seldin, 2002; Keller et al., 2001).

Many other gene products show oscillations similar to the circadian clock. Among them are oncogene proteins NF- κB and tumor suppressor protein p53 and WRN (Nelson et al., 2004; Lahav et al., 2004; Monk, 2003). Such proteins are typically involved in nuclear cytoplasmic shuttling as well as shuttling between different cytoplasmic and nuclear compartments (for instance: between nucleolus and nucleoplasm).

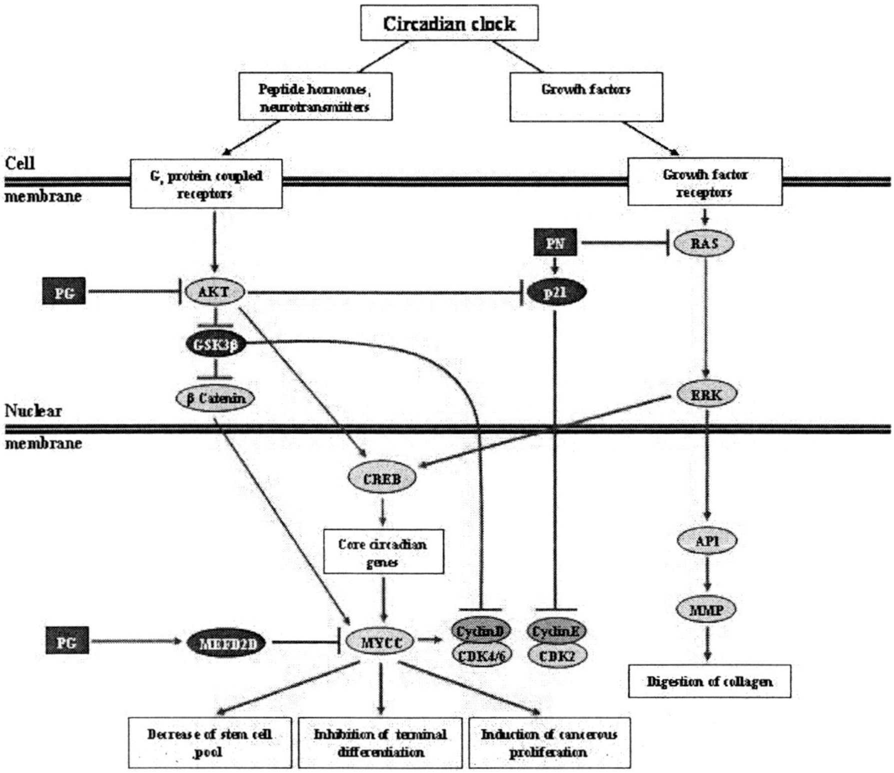

Figure 8. Circadian Clock and Skin Aging. The signal from the central clock regulates cell cycle and apoptosis in peripheral tissue through the release of peptide hormones, neurotransmitters and growth factors, which activate (through Gs protein coupled receptors and growth factor receptors), signal transduction through AKT and RAS pathways. AKT activates β catenin and MYCC and core circadian genes through CREB (which also activates MYCC). This leads to decrease of stem cell pool, inhibition of terminal differentiation and induction of cancerous proliferation. RAS activates MMP through ERK and AP1 and promotes digestion of collagen. PG may reduce these negative effects through decrease of expression of *AKT2*, activation of GSK3β (through inhibition of AKT) and inhibition of MYCC (through MEFD2D) and PN through interruption of signal transduction in RAS pathway and activation of p21.

Circadian Clock and Division Clock

One of the important questions is whether the circadian clock can be separated from the cell division clock. In both systems, there are sequential phases of transcription, translation, protein modification and degradation. In cultured fibroblasts, the circadian clock exists even in the presence of the blockade of cell division (Balsalobre et al., 1998). Based on the studies of Matsuo et al., the circadian clock regulates the cell division clock through WEE1 kinase, which inhibits serine-threonine kinase Cdc2/cyclin B and mitosis (Matsuo et al., 2003). Their research revealed three E boxes which are the binding sites for CLOCK/BMAL1 in the promoter of *WEE1* gene. The entry of the cell to M-phase is positively controlled by the activation of CDC25 phosphatase through active Cdc2-cyclin B complex and inhibited by WEE1 kinase. The activity of Cdc2-cyclin B is controlled by rhythmic changes in concentration of cyclin B. BMAL1/CLOCK activates transcription of the *WEE1* gene. CRY inhibits expression of WEE1 through binding to E box (Matsuo et al., 2003). A number of additional genes active at G1-S and G2-M have E boxes (CLOCK/BMAL1 recognition elements) in their promoters and show circadian oscillations including *Cyclin B1, Cyclin D1, Cyclin D3, CDK4, MYCC* (Grundschober et al., 2001; Fu and Lee, 2002; Matsuo et al., 2003). These data indicate that the circadian clock controls many other important processes. It is suggested that the molecules of the circadian clock are used by other pathways, depending on the type of cell. Deregulation of the circadian rhythm could lead to cancer and promote aging, but not in all cases.

Involvement of Circadian Clock in Cancer and Aging

To summarize, PER2 targets genes, which are important in cancer pathogenesis, including *MYCC*, *Cyclin D1*, *TP53*, *MDM2* and *Cyclin A*. Disruption of the circadian rhythm increases breast cancer risk, has substantial impact on cancer progression, and decreases survival in breast cancer patients (Keith et al., 2001; Sephton et al., 2000). This also promotes aging by decreasing stem cell pool, inhibition of terminal differentiation, induction of damage to collagen and decrease of DNA damage response. Resetting of circadian clock may reduce incidence of cancer and slow down aging and this can be done through dietary restriction and administration of PG (Balsalobre et al., 2000). PG inhibits expression of AKT and downregulate MYCC through MEFD2D (Waldbillig and Burzynski, 2003; Waldbillig et al., 2004). It should help resetting a circadian clock and help increase stem cell pool, terminal differentiation and reduce the danger of cancerous proliferation. **(Figure 8)**

Replicative Clock

Telomeres

Replicative senescence is mediated by cyclin-dependent kinase inhibitors, which lead to cell cycle arrest and stop cell replication (Rohme, 1981). It is caused by telomere shortening, contrary to premature senescence, which is caused by oncogenic stress. The tips of chromosomes in eukaryotic organisms are equipped with telomeres which secure genomic (chromosomal) stability. The loss of telomeres leads to end-to-end fusions of chromosomes, chromosomal abnormalities and cancer. Telomere shorten at each cell cycle, if there are not sufficient quantities of telomerase to maintain them (Bodnar et al., 1996). In stem cells, and in approximately 90% of neoplastic cells, the shortening of telomeres is counteracted by telomerase (Shay and Bacchetti, 1997; Cong et al., 2002).

Telomerase

Telomerase, which is a ribonucleoprotein with a reverse transcriptase activity, adds specific oligonucleotides to the G-rich repeat sequence using an intrinsic RNA template (Nugent and Lundblad, 1998). Telomerase consists of the catalytic protein subunit which is human telomerase reverse transcriptase (hTERT) and human telomerase RNA component (hTERC) (Fanti et al., 1998). hTERC is widely expressed in human cells, but the expression of hTERT is strictly regulated and its level of expression is too low to maintain telomere length (Fanti et al., 1998). Basal keratinocytes, lymphocytes, CD34 expressing peripheral blood stem cells, intestinal crypt cells and immortal cells with alternative lengthening of telomeres (ALT) phenotype possess the telomerase activity or telomerase type activity which is needed for continuous cell renewal (Yasumoto et al., 1996; Counter et al, 1992; 1995; Yui et al., 1998; Hiyama et al., 1996; Bryan et al., 1998). In certain types of immortal cell lines, telomerase is not required for a stable telomere length (Bryan et al., 1995). Such ALT is not yet well characterized but, in yeast, it involves recombination (Lundblad and Szostak, 1989). Short telomeres could promote chromosomal instability and lead to mutations and neoplastic growth.

Transcription of hTERT is regulated by alternative splicing (Villa et al., 2001), methylation of hTERT promoter (Takakura et al., 2001), binding of transcription factors (such as MYCC to hTERT promoter) (Wu et al., 1999), protein kinase C (Li et al., 1998), zinc (Nemoto et al., 2000), calcium (Alfonso-De Matte et al., 2001) and DMSO (Bestilny et al., 1996). Expression of hTERT is regulated through epigenetic mechanisms. Telomeres contain features of constitutive heterochromatin, such as di-and trimethylated H3-K9 and binding of HP1 (Lachner et al., 2001). HP1 is required for telomere capping (Cenci et al., 2003; Fanti et al., 1998). In the absence of HMTs monomethylation of H3-K9 prevails over di-and trimethylation. There is less binding by telomeric chromatin of chromobox proteins Cbx3, Cbx5 and Cbx1 (homologs of *D. melanogaster* heterochromatin protein HP1) and there is loss of heterochromatic features of telomeres and abnormal elongation (Garcia-Cao et al., 2004). This indicates that telomere length can be regulated by methylation.

UV irradiation and DMSO induce telomerase activity in ovarian epithelial cell lines through PI3K (Alfonso-De Matte et al., 2001). c-Jun NH_2-terminal kinase (JNK), which is a target of PI3K, induces telomerase activity, but AKT, which is a primary target for PI3K, is not involved (Alfonso-De Matte et al., 2002). JNK activates c-Jun, triggering transcription of hTERT.

Three pathways; TGF-β, MAD1 and menin negatively control expression of hTERT (Lin and Elledge, 2003). **(Figure 9)**

Telomeres Attrition in Aging

Human telomeres lose approximately 100 base pairs per population doubling, leading to senescence after approximately 50 population doublings (Harley et al. 1990; Counter et al., 1992; Shay and Wright, 2000). The damage to telomeres can not be repaired in most human somatic cells because they do not contain active telomerase (Lachner et al., 2001). Telomere attrition or damage activates ATM-dependent checkpoint response and p53 and p21 pathways leading to cell cycle arrest and apoptosis (van Steensel et al., 1998; d'Adda di Fagagna et al., 2003). ROS are major factors causing oxidative damage to telomeres (Huffman et al., 2000). Another factor which plays a major part in this type of senescence is activation of a CDK4/6 inhibitor p16 (Bond et al., 2004).

Telomere Insufficiency in Progeroid Syndrome

Tumor-suppressor gene *WRN* plays an important part in telomere maintenance. Loss of function of *WRN* gene is responsible for WS which is an autosomal recessive genetic disorder (Kipling et al., 2004). Patients with WS are usually symptom-free during the first 10 years of their life. After that, they develop thinning and wrinkling of the skin, thinning and graying of hair, shorter than normal stature and numerous additional signs of premature aging. They develop neoplasms of mesenchymal origin, soft tissue and osteogenic sarcomas, malignant melanoma and thyroid cancers. They usually do not survive longer than 50 years and die from cardiovascular disease or cancer (Kipling et al., 2004). The senescent fibroblasts in WS are rescued by enforced expression of telomeres (Wyllie et al., 2000). WRN-deficient mice with long telomeres are free from WS (Lombard et al., 2000). The delayed manifestation of WS, after the first decade of life, is possibly due to reduced telomere reserves (Chang et al., 2004). Patients with WS show severe pathological changes in the dermal portions of the skin (Goto et al., 1985). It was found that fibroblasts in WS have markedly lowered capacity for proliferation. Most fibroblasts in WS (90%) have an in vitro life-span of less than 20 population doublings, which compares to a normal 40 to 100 (Salk et al., 1981; Kipling et al., 2004). This resulted in a very rapid decrease of mitotic fraction per population doubling (Kill et al., 1994). The cells have premature replicative senescence and mutator phenotype (Kipling et al., 2004).

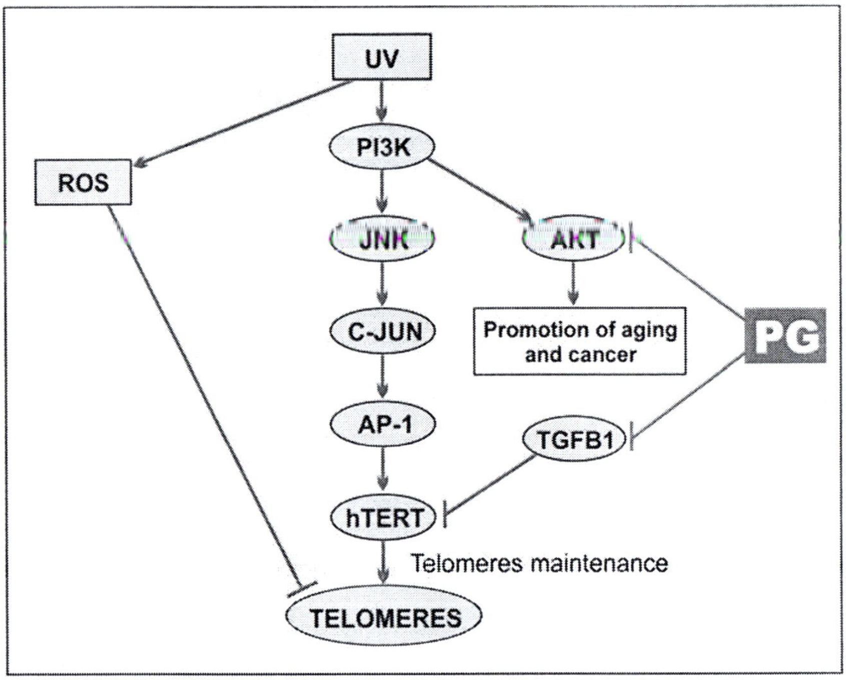

Figure 9. Replicative Clock and Skin Aging. UV irradiation promotes skin aging and cancer through activation of AKT. It also activates hTERT through PI3K. ROS generated through UV induce damage to telomeres and promote aging. PG decreases expression of *AKT2* and may slow down aging and protect against cancer. Decrease of *TGFB1* expression by PG without inhibitory effect on PI3K does not interfere with telomere maintenance.

WRN protein belongs to the RecQ family of helicases (Nakayama et al., 1984) and this group of enzymes is unwinding complimentary strands of nucleic-acid duplexes. The hydrolysis of ATP provides the energy for the reaction (Soultanas and Wigley, 2001). Humans express five enzymes from this group (Hickson, 2003). *WRN* locates to chromosome 8 p-12. In WS, there seem to be nonsense, frameshift and splicing mutations (Hickson, 2003). Since telomeres contain G rich sequences, WRN may assist in telomere lengthening by disrupting G-quadruplexes and may help in ALT. The absence of WRN leads to fusion and breakage of chromosomes. In normal human cells, WRN is localized in nucleoli (Marciniak et al., 1998). It transfers to DNA replication sites when DNA synthesis is blocked (Constantinou et al., 2000). The primary activity of WRN takes place at sites of disrupted synthesis of DNA. The most important proteins, which bind to WRN, include p53, replication protein A (RPA) and proliferating cells nuclear antigen (PCNA) (Wang et al., 2001; Brosh et al., 1999; Lebel et al., 1999). Binding with p53 effects exonuclease and helicase activity of WRN and is necessary for WRN action (Brosh et al., 1999; Sommers et al., 2005). p53-dependant apoptosis is attenuated in WS cells (Spillare et al., 1999).

Possible Effect of Amino Acid Derivatives on Replicative Clock

In progeroid syndrome WS and in chronological aging, the maintenance of telomeres is less effective than in the young age. The telomeres are shorter and there is from 2 to 5 times less population doublings of WS fibroblast cultures compared to normal cells (Salk et al., 1981). Since hTERT is regulated through methylation, it is possible that there is silencing of telomerase in aging. ROS and UV may damage telomeres and TGFB1 may decrease expression of hTERT, further reducing the renewal of the tissues. PG may help to normalize this process through reduction of expression of TGFB1. During exposure to UV light, activation of PI3K increases expression of hTERT. This pathway, however, is separated from PI3K/AKT. This means that decrease of AKT expression by PG should not interfere with stimulation of hTERT expression by PI3K. **(Figure 9)**

Age Management Measures which Target Silenced Genes

Background

Many genes are silenced during aging, but a smaller number of genes have increased activity. The agents, which interrupt signal transmission in IGF1/AKT and RAS pathways, lead to normal expression of tumor suppressor genes *TP53 and p21*. This may help to slow down the aging process and reduce the incidence of cancer. The use of combination of agents, which individually have only moderate activity, is preferable over single potent pharmaceuticals. In this respect, dietary supplements and cosmetics may play an important part. There is no doubt that proper diet and physical activity reduces the incidence of cancer and prolongs life. The advisers to the ruling class, who were priests in Egypt and Mesopotamia, knew about positive effects of the proper nutrition on life extension. At the time when the average life expectancy in population was approximately 30 years the members of the ruling class lived substantially longer. Ramses II survived 92 years and

Pharaoh Neferkare ruled for 94 years (Burzynski, 2005). One of the reasons of such long survival of these rulers was a special diet. This was documented in old scriptures and confirmed by chemical analysis of archeological findings.

Scientifically rational anti-aging intervention should attempt to change expression of the genes to the level typical for a young adult age. Silenced genes, such as those involved in detoxification, cholesterol metabolism and tumor suppressors, should be activated. The excessive expression of a different group of genes should be brought to normal, including those involved with chronic inflammation, stress response and protection from apoptosis (Burzynski, 2003; 2004b, 2005). Dietary restriction and increased physical activity is a simple and proven method which should be tried first. The studies in animals and humans confirmed that interruption of signal transmission along IGF1 pathway extends life and reduces cancer incidence (Kenyon, 2001; Cao et al., 2001). Numerous dietary ingredients have been recommended for prevention of cancer and they can also be applied in age management. The list of the supplements is very extensive and is growing everyday (Surh, 2003). Good examples are curcumin from turmeric, resveratrol from grapes and limonene from citrus, which interrupt signal transmission in RAS pathway, inhibit NF-κB, AP1 and COX2 and mediate degradation of β catenin; however, the negative effect of resveratrol is inactivation of p53. Epigallocatechin-3-gallate (EGCG) from green tea and genistein decrease signaling in AKT pathway. For extensive description of these phytochemicals, the reader is directed to excellent review articles (Surh, 2003).

A10, PG, isoPG, PN and PG, occur naturally in human plasma and in dairy products and vegetables (Burzynski, 2004b, 2001; Burzynski et al., 2003). They influence the genes involved in formation of cancer and in aging. PG effects gene expression of a number of genes which was described earlier in this chapter. From a perspective of application of PG in antiaging regimen, the most important is decreased expression of *AKT2, TGFB1* and increased activity of *PTEN* (Waldbillig and Burzynski, 2003; Waldbillig et al., 2004). The use of PG should restore proper balance in the activity of important genes involved in aging, such as *IGF1/AKT, TGFB1* and *PTEN*. PG and isoPG are ingredients of cosmetic products; Aminocare® cream and Aminocare® lotion, which were designed to help maintain healthier and younger looking skin, reduce the wrinkles and lessen, or eliminate, skin discoloration. Formation of wrinkles may involve increased signaling through IGF1/AKT2 pathway, increased activity of TGFB1 and downregulation of PTEN which can be normalized by PG.

Clinical Study in Skin Aging

Aminocare® facial cream was submitted for clinical study to evaluate the anti-wrinkle effect on the skin (Burzynski et al., 2004). The study tested tolerance (occlusive patch test), anti-wrinkle efficacy and acceptability of the cream containing, correspondingly, 0.5% and 0.1% of synthetic PG and isoPG. It was performed at the Department of Pharmacology of the University Victor Segalen in Bordeaux, France and was supervised by the Ethics Committee of the university. The volunteers who were 22 healthy women from 41 to 61 years of age were recruited from the university personnel. One of the volunteers did not appear for evaluation at the end of the study; therefore, the final results were reported on 21 valid observations.

The evaluation of the tolerance of Aminocare® cream began 3 days before the study of anti-wrinkle effectiveness. After explanation of the details of the study, the volunteers read, understood and signed the informed consent form. The history was taken and a physical examination was performed, followed by application of the patch test (Finn Chamber ®) on the area of the back. The patch was removed 48 hours later. The test was read after 15 minutes and the volunteers were questioned about possible undesireable effects. After 72 hours, the volunteers were questioned again followed by repeated reading of the test. The following undesirable events, if present, would be graded on a scale from 0 (none) to 4 (serious): erythema, edema, blisters, dryness, roughness and reflexivity.

For the purpose of evaluation of anti-wrinkle efficacy, the cream was applied twice a day to the entire face for 28 days, except for one of the areas of "crow's feet" wrinkles, which served as the "non-treated" area. The imprints of the wrinkles of "crows feet" of the treated and non-treated areas were produced by using silicone Elite H-D+ on day-0 and day-28. The volunteers were given instructions and questionnaires to be completed at the end of the study. The history and physical examinations were performed again on day-28 followed by evaluation of questionnaires. Anti-wrinkle efficacy was evaluated through objective laser profilometry of the imprints with data submitted for statistical analysis by using Student t and Wilcoxon tests. The measurements of the volume of wrinkles, depth and complexity at the beginning and at the end of the study were compared to the controls. The volunteers expressed their opinion of the product in the questionnaires at the end of the study and the investigator rendered his opinion after taking a history and performing a physical examination.

There were no reactions detected on the sites where patches were applied. Two volunteers reported mild and moderate symptoms during the study, but there were no further consequences. These observations indicated that tolerance of Aminocare® facial cream was excellent. The results of statistical analysis of the profilometric images showed that the treated and non-treated sites were comparable at the start of the study. All three variables, complexity, depth and volume had each significantly decreased on the sites treated with Aminocare® cream at the end of the study and there were neither changes nor continued expansion on the non-treated sites. Based on these results, anti-wrinkle efficacy was completely satisfactory. **(Figure 10)** The study volunteers expressed many favorable comments about Aminocare® cream. They appreciated the softness, suppleness and additional skin comfort they noticed after the first Aminocare® cream application. With the exception of two women, all volunteers expressed a definite favorable desire to continue using Aminocare® cream.

The studies in human volunteers with Aminocare® facial creams proved the antiaging effect, which theoretically is based on regulation of altered gene expression in aging by PG. This is a preliminary study and did not include histopathological examination of skin specimen before and after application of the cream to confirm that the proposed effect on the genes involved in aging takes place. Such additional studies can be performed in the future as well as clinical trials with oral dosing of PG to objectively test additional parameters of aging.

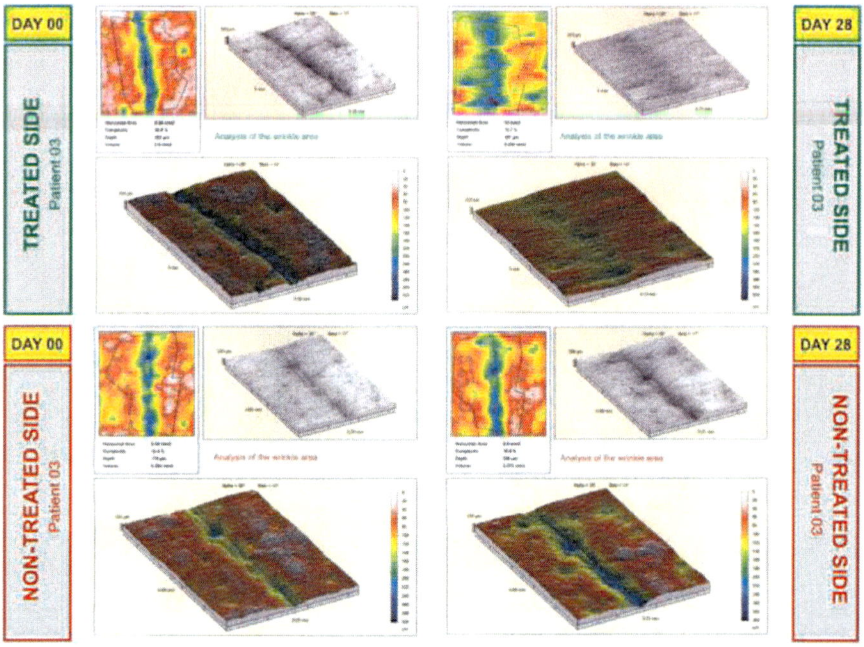

Figure 10. Laser Profilometry of Skin Treated with Aminocare® Cream Compared to Control. Silicone imprints of the treated and non-treated areas of complex wrinkles were taken at the beginning and end of the study. The imprints were examined by laser profilometry followed by statistical analysis to measure changes in the complexity, depth and volume of wrinkles. The images clearly show positive changes and the reduction of wrinkles in the treated area, and the continued expansion of wrinkles in the non-treated area for study participant No. 03.

Concluding Remarks and Perspectives

The results of the studies by numerous research teams indicated substantial changes in gene expression in aging process with most of the genes showing decreased expression. Epigenetic activities, which include hypermethylation of promoters regions, global hypomethylation, chromatin remodeling and histone modification, are among mechanisms responsible for changes of gene expression in aging. These mechanisms are reversible and can be modulated by pharmaceutical agents, diet and dietary supplements and even cosmetics generating the hope that such measures may slow down the aging process. The derivatives of glutamine, phenylacetate and phenylbutyrate, which occur naturally in human blood and urine, dairy products and vegetables, may help to activate silenced tumor suppressor genes, *TP53, p21, PTEN* and *INI1* and decrease overexpression of oncogenes, *RAS, AKT* and *MYCC*.

Aging of the skin is of major concern and is influenced by external factors, changes in gene expression, circadian clock and replicative clock. Theoretically, and based on the research, a simple amino acid derivative PG may, through epigenetic mechanisms, normalize expression of some important genes involved in aging and reset biological clocks. Preliminary evidence of this possibility is provided by clinical study in skin aging where a cosmetic cream containing PG caused statistically significant decrease of wrinkles which are the major sign of aging. It should be stressed that these are preliminary results and as such should be treated with caution. Identification of dose- dependent changes of gene expression by PG is only the first step in a complex process leading to elucidation of translational, posttranslational and metabolic changes of gene products. Each of these important processes may change the action of PG. Gene array studies were performed on human glioma U-87 cells and it is entirely possible that the effect on cells involved in skin aging will be quite different. The volunteers did not have histopathological examination of the skin before and after application of the cream and the anti-wrinkle effect was determined only by laser profilometry. On the positive side, the results of the study are encouraging and the volunteers did not suffer adverse reactions. Future studies can address and resolve the issues of concern.

The users of oral dietary supplement, Aminocare® A10, reported a number of positive antiaging effects: reduction of wrinkles and hyperpigmentation spots, reduction of serum cholesterol, decrease of frequency of common viral infections, improvement of chronic inflammation, antidepressant effect, increased energy and alleviation of symptoms of benign prostate hypertrophy. These are, however, anecdotal observations and some of them will be verified by clinical studies. It is hoped that future research on normalization of changes of gene expression in aging will deliver important results which can be readily applied in age management practice.

Acknowledgements

The author gratefully acknowledges Tomasz Janicki, M.D., Elaine Dittoe and Adam Golunski of Burzynski Clinic for editorial assistance.

References

Adam L, Crepin M, Savin C, Israel L. Sodium phenylacetate induces growth inhibition and Bcl-2 down-regulation and apoptosis in MCF7ras cells in vitro and in nude mice. *Cancer Res* 1995;55:5156-60.

Adam L, Crepin M, Israel L. Tumor growth inhibition, apoptosis, and Bcl-2 down-regulation of MCF-7ras tumors by sodium phenylacetate and tamoxifen combination. *Cancer Res* 1997;57:1023-9.

Ahuja N, Li Q, Mohan AL, Baylin SB, Issa JP. Aging and DNA methylation in colorectal mucosa and cancer. *Cancer Res* 1998;58:5489-94.

Akiyama Y, Maesawa C, Ogasawara S, Terashima M, Masuda T. Cell-type-specific repression of the maspin gene is disrupted frequently by demethylation at the promoter region in gastric intestinal metaplasia and cancer cells. *Am J Pathol* 2003;163:1911-9.

Alfonso-De Matte MY, Cheng JQ, Kruk PA. Ultraviolet irradiation- and dimethyl sulfoxide-induced telomerase activity in ovarian epithelial cell lines. Exp Cell Res 2001;267:13-27.

Alves G, Tatro A, Fanning T. Differential methylation of human LINE-1 retrotransposons in malignant cells. *Gene* 1996;176:39-44.

Ashraf AQ, Liau MC, Mohabbat MO, Burzynski SR. Preclinical studies on antineoplaston A10 injections. *Drugs Exp Clin Res* 1986;12 Suppl 1:37-45.

Ayroldi E, Zollo O, et al. Glucocorticoid-induced leucine zipper inhibits the Raf-extracellular signal-regulated kinase pathway by binding to Raf-1. *Mol Cell Biol* 2002;22:7929-41.

Balsalobre A, Damiola F, Schibler U. A serum shock induces circadian gene expression in mammalian tissue culture cells. *Cell* 1998;93:929-37.

Balsalobre A, Marcacci L, Schibler U. Multiple signaling pathways elicit circadian gene expression in cultured Rat-1 fibroblasts. *Curr Biol* 2000;10:1291-4.

Bannister AJ, Zegerman P, et al. Selective recognition of methylated lysine 9 on histone H3 by the HP1 chromo domain. *Nature* 2001;410:120-4.

Bednar J, Horowitz RA, et al. Nucleosomes, linker DNA, and linker histone form a unique structural motif that directs the higher-order folding and compaction of chromatin. *Proc Natl Acad Sci U S A* 1998;95:14173-8.

Bestilny LJ, Brown CB, Miura Y, Robertson LD, Riabowol KT. Selective inhibition of telomerase activity during terminal differentiation of immortal cell lines. *Cancer Res* 1996;56:3796-802.

Betz BL, Strobeck MW, Reisman DN, Knudsen ES, Weissman BE. Re-expression of hSNF5/INI1/BAF47 in pediatric tumor cells leads to G1 arrest associated with induction of p16ink4a and activation of RB. *Oncogene* 2002;21:5193-203.

Bhattacharya SK, Ramchandani S, Cervoni N, Szyf M. A mammalian protein with specific demethylase activity for mCpG DNA. *Nature* 1999;397:579-83.

Bodnar AG, Kim NW, Effros RB, Chiu CP. Mechanism of telomerase induction during T cell activation. *Exp Cell Res* 1996;228:58-64.

Bond J, Jones C, et al. Direct evidence from siRNA-directed "knock down" that p16(INK4a) is required for human fibroblast senescence and for limiting ras-induced epithelial cell proliferation. *Exp Cell Res* 2004;292:151-6.

Boudreault AA, Cronier D, et al. Yeast enhancer of polycomb defines global Esa1-dependent acetylation of chromatin. *Genes Dev* 2003;17:1415-28.

Bowden GT. Prevention of non-melanoma skin cancer by targeting ultraviolet-B-light signalling. *Nat Rev Cancer* 2004;4:23-35.

Brosh RM, Jr., Orren DK, et al. Functional and physical interaction between WRN helicase and human replication protein A. *J Biol Chem* 1999;274:18341-50.

Brosh RM, Jr., Karmakar P, et al. p53 Modulates the exonuclease activity of Werner syndrome protein. *J Biol Chem* 2001;276:35093-102.

Brown K, Strathdee D, Bryson S, Lambie W, Balmain A. The malignant capacity of skin tumours induced by expression of a mutant H-ras transgene depends on the cell type targeted. *Curr Biol* 1998;8:516-24.

Bryan TM, Englezou A, Gupta J, Bacchetti S, Reddel RR. Telomere elongation in immortal human cells without detectable telomerase activity. *Embo J* 1995;14:4240-8.

Bryan TM, Englezou A, Dunham MA, Reddel RR. Telomere length dynamics in telomerase-positive immortal human cell populations. *Exp Cell Res* 1998;239:370-8.

Bulger M, Sawado T, Schubeler D, Groudine M. ChIPs of the beta-globin locus: unraveling gene regulation within an active domain. *Curr Opin Genet Dev* 2002;12:170-7.

Burzynski SR. Gene silencing--a new theory of aging. Med Hypotheses 2003;60:578-83.

Burzynski SR. The present state of antineoplaston research (1). *Integr Cancer Ther* 2004a;3:47-58.

Burzynski SR. Mechanisms and prevention of genetic aging. *Neurologia i Psychiatria* 2004b;4:1-8.

Burzynski SR. Aging: gene silencing or gene activation? *Med Hypotheses* 2005;64:201-8.

Burzynski SR, Hendry LB, Mohabbat MO, Liau MC, Khalid M, Burzynski B. Purification, structure determination, synthesis and animal toxicity studies of Antineoplaston A10. *Proc 13th Int Cong Chemother* 1983;17:12.4.11-.4.14.

Burzynski SR, Mohabbat MO, Burzynski B. Human toxicology studies on oral formulation of Antineoplaston A10. *Drugs Exp Clin Res* 1984;10:891-909.

Burzynski SR, Mohabbat MO, Lee SS. Preclinical studies on antineoplaston AS2-1 and antineoplaston AS2-5. *Drugs Exp Clin Res* 1986;12 Suppl 1:11-6.

Burzynski SR, Jacquet A, Moore N, Musial L. Gene silencing theory of aging: The clinical trial supporting the theory. Presented at the 12th Annual World Congress on Anti-Aging Medicine. *American Academy of Anti-Aging Medicine* (A4M), Las Vegas 2004.

Burzynski SR, Ilkowska-Musial E, Klimczak MW, Musial L. Antineoplastons in Dairy Products. *J Appl Nutr* (In Press) 2005.

Buscher M, Rahmsdorf HJ, Litfin M, Karin M, Herrlich P. Activation of the c-fos gene by UV and phorbol ester: different signal transduction pathways converge to the same enhancer element. *Oncogene* 1988;3:301-11.

Cai S, Han HJ, Kohwi-Shigematsu T. Tissue-specific nuclear architecture and gene expression regulated by SATB1. *Nat Genet* 2003;34:42-51.

Cao SX, Dhahbi JM, Mote PL, Spindler SR. Genomic profiling of short- and long-term caloric restriction effects in the liver of aging mice. *Proc Natl Acad Sci U S A* 2001;98:10630-5.

Castillo M, Iglesias J, Zafra MF, Garcia-Peregrin E. Effect of phenylpyruvate on mevalonate-activating enzymes from chick brain and liver. *Arch Int Physiol Biochim Biophys* 1991;99:165-8.

Cenci G, Siriaco G, Raffa GD, Kellum R, Gatti M. The Drosophila HOAP protein is required for telomere capping. *Nat Cell Biol* 2003;5:82-4.

Chan HM, La Thangue NB. p300/CBP proteins: HATs for transcriptional bridges and scaffolds. *J Cell Sci* 2001;114:2363-73.

Chang CC, Kuo IC, et al. Detection of quadruplex DNA structures in human telomeres by a fluorescent carbazole derivative. *Anal Chem* 2004;76:4490-4.

Channavajhala P, Seldin DC. Functional interaction of protein kinase CK2 and c-Myc in lymphomagenesis. *Oncogene* 2002;21:5280-8.

Chung JH, Seo JY, et al. Modulation of skin collagen metabolism in aged and photoaged human skin in vivo. *J Invest Dermatol* 2001;117:1218-24.

Collins AF, Pearson HA, et al. Oral sodium phenylbutyrate therapy in homozygous beta thalassemia: a clinical trial. *Blood* 1995;85:43-9.

Cong YS, Wright WE, Shay JW. Human telomerase and its regulation. *Microbiol Mol Biol Rev* 2002;66:407-25, table of contents.

Constantinou A, Tarsounas M, et al. Werner's syndrome protein (WRN) migrates Holliday junctions and co-localizes with RPA upon replication arrest. *EMBO Rep* 2000;1:80-4.

Costello JF, Plass C. Methylation matters. *J Med Genet* 2001;38:285-303.

Counter CM, Avilion AA, et al. Telomere shortening associated with chromosome instability is arrested in immortal cells which express telomerase activity. *Embo J* 1992;11:1921-9.

Counter CM, Gupta J, Harley CB, Leber B, Bacchetti S. Telomerase activity in normal leukocytes and in hematologic malignancies. *Blood* 1995;85:2315-20.

d'Adda di Fagagna F, Reaper PM, et al. A DNA damage checkpoint response in telomere-initiated senescence. *Nature* 2003;426:194-8.

Damiola F, Le Minh N, et al. Restricted feeding uncouples circadian oscillators in peripheral tissues from the central pacemaker in the suprachiasmatic nucleus. *Genes Dev* 2000;14:2950-61.

DiDonato JA, Hayakawa M, Rothwarf DM, Zandi E, Karin M. A cytokine-responsive IkappaB kinase that activates the transcription factor NF-kappaB. *Nature* 1997;388:548-54.

Dobrovic A, Simpfendorfer D. Methylation of the BRCA1 gene in sporadic breast cancer. *Cancer Res* 1997;57:3347-50.

Dorigo B, Schalch T, et al. Nucleosome arrays reveal the two-start organization of the chromatin fiber. *Science* 2004;306:1571-3.

Ducker CE, Simpson RT. The organized chromatin domain of the repressed yeast a cell-specific gene STE6 contains two molecules of the corepressor Tup1p per nucleosome. *Embo J* 2000;19:400-9.

Ducut Sigala JL, Bottero V, et al. Activation of transcription factor NF-kappaB requires ELKS, an IkappaB kinase regulatory subunit. *Science* 2004;304:1963-7.

Ehrlich M. DNA methylation in cancer: too much, but also too little. *Oncogene* 2002;21:5400-13.

Eriguchi N, Hara H, et al. Chemopreventive effect of antineoplaston A-10 on urethane-induced pulmonary neoplasm in mice. *Nippon Gan Chiryo Gakkai Shi* 1988;23:1560-5.

Esteller M. Dormant hypermethylated tumour suppressor genes: questions and answers. *J Pathol* 2005;205:172-80.

Etchegaray JP, Lee C, Wade PA, Reppert SM. Rhythmic histone acetylation underlies transcription in the mammalian circadian clock. *Nature* 2003;421:177-82.

Fan T, Yan Q, et al. Lsh-deficient murine embryonal fibroblasts show reduced proliferation with signs of abnormal mitosis. *Cancer Res* 2003;63:4677-83.

Fanti L, Giovinazzo G, Berloco M, Pimpinelli S. The heterochromatin protein 1 prevents telomere fusions in Drosophila. *Mol Cell* 1998;2:527-38.

Feinberg AP, Tycko B. The history of cancer epigenetics. *Nat Rev Cancer* 2004;4:143-53.

Fisher GJ, Wang ZQ, et al. Pathophysiology of premature skin aging induced by ultraviolet light. *N Engl J Med* 1997;337:1419-28.

Flanagan JF, Peterson CL. A role for the yeast SWI/SNF complex in DNA replication. *Nucleic Acids Res* 1999;27:2022-8.

Francis NJ, Saurin AJ, Shao Z, Kingston RE. Reconstitution of a functional core polycomb repressive complex. *Mol Cell 2001*;8:545-56.

Francis NJ, Kingston RE, Woodcock CL. Chromatin compaction by a polycomb group protein complex. *Science* 2004;306:1574-7.

Fremont M, Siegmann M, et al. Demethylation of DNA by purified chick embryo 5-methylcytosine-DNA glycosylase requires both protein and RNA. *Nucleic Acids Res* 1997;25:2375-80.

Fry CJ, Peterson CL. Chromatin remodeling enzymes: who's on first? *Curr Biol* 2001;11:R185-97.

Fu L, Pelicano H, Liu J, Huang P, Lee C. The circadian gene Period2 plays an important role in tumor suppression and DNA damage response in vivo. *Cell* 2002;111:41-50.

Fu L, Lee CC. The circadian clock: pacemaker and tumour suppressor. *Nat Rev Cancer* 2003;3:350-61.

Gama-Sosa MA, Slagel VA, et al. The 5-methylcytosine content of DNA from human tumors. *Nucleic Acids Res* 1983;11:6883-94.

Garcia MJ, Martinez-Delgado B, et al. Different incidence and pattern of p15INK4b and p16INK4a promoter region hypermethylation in Hodgkin's and CD30-Positive non-Hodgkin's lymphomas. *Am J Pathol* 2002;161:1007-13.

Garcia-Cao M, O'Sullivan R, Peters AH, Jenuwein T, Blasco MA. Epigenetic regulation of telomere length in mammalian cells by the Suv39h1 and Suv39h2 histone methyltransferases. *Nat Genet* 2004;36:94-9.

Geigl JB, Langer S, et al. Analysis of gene expression patterns and chromosomal changes associated with aging. *Cancer Res* 2004;64:8550-7.

Gekakis N, Staknis D, et al. Role of the CLOCK protein in the mammalian circadian mechanism. *Science* 1998;280:1564-9.

Georgel PT, Horowitz-Scherer RA, et al. Chromatin compaction by human MeCP2. Assembly of novel secondary chromatin structures in the absence of DNA methylation. J *Biol Chem* 2003;278:32181-8.

Gessi M, Giangaspero F, Pietsch T. Atypical teratoid/rhabdoid tumors and choroid plexus tumors: when genetics "surprise" pathology. *Brain Pathol* 2003;13:409-14.

Ghazizadeh S, Taichman LB. Multiple classes of stem cells in cutaneous epithelium: a lineage analysis of adult mouse skin. *Embo J* 2001;20:1215-22.

Ghosh S, May MJ, Kopp EB. NF-kappa B and Rel proteins: evolutionarily conserved mediators of immune responses. *Annu Rev Immunol* 1998;16:225-60.

Giampapa VC. Anti-aging medicine and cosmetic surgery "the connection". 2003. In: Giampapa VC (Ed), The Basic Principles and Practice of Anti-Aging Medicine & Age Management for the Aesthetic Surgeon and Physician. Newark: *Giampapa* 2003:1-14.

Gibbons RJ, McDowell TL, et al. Mutations in ATRX, encoding a SWI/SNF-like protein, cause diverse changes in the pattern of DNA methylation. *Nat Genet* 2000;24:368-71.

Goelz SE, Vogelstein B, Hamilton SR, Feinberg AP. Hypomethylation of DNA from benign and malignant human colon neoplasms. *Science* 1985;228:187-90.

Gonzalez-Gomez P, Bello MJ, et al. CpG island methylation status and mutation analysis of the RB1 gene essential promoter region and protein-binding pocket domain in nervous system tumours. *Br J Cancer* 2003;88:109-14.

Gorospe M, Shack S, Guyton KZ, Samid D, Holbrook NJ. Up-regulation and functional role of p21Waf1/Cip1 during growth arrest of human breast carcinoma MCF-7 cells by phenylacetate. *Cell Growth Differ* 1996;7:1609-15.

Goto M, Tanimoto K, Miyamoto T. Immunological aspects of Werner's syndrome: an analysis of 17 patients. *Adv Exp Med Biol* 1985;190:263-84.

Grundschober C, Delaunay F, et al. Circadian regulation of diverse gene products revealed by mRNA expression profiling of synchronized fibroblasts. *J Biol Chem* 2001;276:46751-8.

Guha A, Feldkamp MM, Lau N, Boss G, Pawson A. Proliferation of human malignant astrocytomas is dependent on Ras activation. *Oncogene* 1997;15:2755-65.

Hagihara A, Miyamoto K, et al. Identification of 27 5' CpG islands aberrantly methylated and 13 genes silenced in human pancreatic cancers. *Oncogene* 2004;23:8705-10.

Harden SV, Tokumaru Y, et al. Gene promoter hypermethylation in tumors and lymph nodes of stage I lung cancer patients. *Clin Cancer Res* 2003;9:1370-5.

Harley CB, Futcher AB, Greider CW. Telomeres shorten during ageing of human fibroblasts. *Nature* 1990;345:458-60.

Hegi ME, Diserens AC, et al. Clinical trial substantiates the predictive value of O-6-methylguanine-DNA methyltransferase promoter methylation in glioblastoma patients treated with temozolomide. *Clin Cancer Res* 2004;10:1871-4.

Hendry LB, Muldoon TG, Burzynski SR, Copland JA, Lehner AF. Stereochemical modelling studies of the interaction of antineoplaston A10 with DNA. *Drugs Exp Clin Res* 1987;13 Suppl 1:77-81.

Hendry LB, Muldoon TG. Actions of an endogenous antitumorigenic agent on mammary tumor development and modeling analysis of its capacity for interacting with DNA. *J Steroid Biochem* 1988;30:325-8.

Herman JG, Merlo A, et al. Inactivation of the CDKN2/p16/MTS1 gene is frequently associated with aberrant DNA methylation in all common human cancers. *Cancer Res* 1995;55:4525-30.

Herms J, Neidt I, et al. C-MYC expression in medulloblastoma and its prognostic value. *Int J Cancer* 2000;89:395-402.

Herrlich P. Cross-talk between glucocorticoid receptor and AP-1. *Oncogene* 2001;20:2465-75.

Hickson ID. RecQ helicases: caretakers of the genome. *Nat Rev Cancer* 2003;3:169-78.

Hirschhorn JN, Brown SA, Clark CD, Winston F. Evidence that SNF2/SWI2 and SNF5 activate transcription in yeast by altering chromatin structure. *Genes Dev* 1992;6:2288-98.

Hiyama E, Hiyama K, Tatsumoto N. Telomerase activity in human intestine. *Int J Oncol* 1996;9:453-458.

Hoal-van Helden EG, van Helden PD. Age-related methylation changes in DNA may reflect the proliferative potential of organs. *Mutat Res* 1989;219:263-6.

Hogenesch JB, Gu YZ, Jain S, Bradfield CA. The basic-helix-loop-helix-PAS orphan MOP3 forms transcriptionally active complexes with circadian and hypoxia factors. *Proc Natl Acad Sci U S A* 1998;95:5474-9.

Horn PJ, Peterson CL. Molecular biology. Chromatin higher order folding--wrapping up transcription. *Science* 2002;297:1824-7.

Howitz KT, Bitterman KJ, et al. Small molecule activators of sirtuins extend Saccharomyces cerevisiae lifespan. *Nature* 2003;425:191-6.

Hsu DS, Zhao X, et al. Putative human blue-light photoreceptors hCRY1 and hCRY2 are flavoproteins. *Biochemistry* 1996;35:13871-7.

Hsu HL, Gilley D, Blackburn EH, Chen DJ. Ku is associated with the telomere in mammals. *Proc Natl Acad Sci U S A* 1999;96:12454-8.

Hu MC, Lee DF, et al. IkappaB kinase promotes tumorigenesis through inhibition of forkhead FOXO3a. *Cell* 2004;117:225-37.

Huang S. Histone methyltransferases, diet nutrients and tumour suppressors. *Nat Rev Cancer* 2002;2:469-76.

Huffman KE, Levene SD, Tesmer VM, Shay JW, Wright WE. Telomere shortening is proportional to the size of the G-rich telomeric 3'-overhang. *J Biol Chem* 2000;275:19719-22.

Hunter JA, Savin JA, Dahl MV. Clinical Dermatology (Blackwell Science, Oxford, 1995).

Issa JP. Epigenetic variation and human disease. *J Nutr* 2002;132:2388S-2392S.

Issa JP. CpG island methylator phenotype in cancer. *Nat Rev Cancer* 2004;4:988-93.

Issa JP, Ottaviano YL, et al. Methylation of the oestrogen receptor CpG island links ageing and neoplasia in human colon. *Nat Genet* 1994;7:536-40.

Issa JP, Ahuja N, Toyota M, Bronner MP, Brentnall TA. Accelerated age-related CpG island methylation in ulcerative colitis. *Cancer Res* 2001;61:3573-7.

Jaenisch R, Bird A. Epigenetic regulation of gene expression: how the genome integrates intrinsic and environmental signals. *Nat Genet* 2003;33 Suppl:245-54.

James MO, Smith RL, Williams RT, Reidenberg M. The conjugation of phenylacetic acid in man, sub-human primates and some non-primate species. *Proc R Soc Lond B Biol Sci* 1972;182:25-35.

Jenuwein T, Allis CD. Translating the histone code. *Science* 2001;293:1074-80.

Johnson GL, Lapadat R. Mitogen-activated protein kinase pathways mediated by ERK, JNK, and p38 protein kinases. *Science* 2002;298:1911-2.

Jurgens B, Schmitz-Drager BJ, Schulz WA. Hypomethylation of L1 LINE sequences prevailing in human urothelial carcinoma. *Cancer Res* 1996;56:5698-703.

Kamitani H, Taniura S, et al. Histone acetylation may suppress human glioma cell proliferation when p21 WAF/Cip1 and gelsolin are induced. *Neuro-oncol* 2002;4:95-101.

Kampalath BN, Liau MC, Burzynski B, Burzynski SR. Chemoprevention by antineoplaston A10 of benzo(a)pyrene-induced pulmonary neoplasia. *Drugs Exp Clin Res* 1987;13 Suppl 1:51-5.

Kampalath BN, Liau MC, Burzynski SR, Burzynski SR. Protective effect of antineoplastin A10 in hepatocarcinogenesis induced by aflatoxin B1. *Int J Tissue React* 1990;12 (1):43-50.

Kaneda A, Kaminishi M, Yanagihara K, Sugimura T, Ushijima T. Identification of silencing of nine genes in human gastric cancers. *Cancer Res* 2002;62:6645-50.

Kang GH, Lee S, et al. Epstein-barr virus-positive gastric carcinoma demonstrates frequent aberrant methylation of multiple genes and constitutes CpG island methylator phenotype-positive gastric carcinoma. *Am J Pathol* 2002;160:787-94.

Karlsson S, Nienhuis AW. Developmental regulation of human globin genes. *Annu Rev Biochem* 1985;54:1071-108.

Kawara S, Mydlarski R, et al. Low-dose ultraviolet B rays alter the mRNA expression of the circadian clock genes in cultured human keratinocytes. *J Invest Dermatol* 2002;119:1220-3.

Keith LG, Oleszczuk JJ, Laguens M. Circadian rhythm chaos: a new breast cancer marker. *Int J Fertil Womens Med* 2001;46:238-47.

Keller DM, Zeng X, et al. A DNA damage-induced p53 serine 392 kinase complex contains CK2, hSpt16, and SSRP1. *Mol Cell* 2001;7:283-92.

Kenyon C. A conserved regulatory system for aging. Cell 2001;105:165-8.

Kill IR, Faragher RG, Lawrence K, Shall S. The expression of proliferation-dependent antigens during the lifespan of normal and progeroid human fibroblasts in culture. *J Cell Sci* 1994;107 (Pt 2):571-9.

King DP, Zhao Y, et al. Positional cloning of the mouse circadian clock gene. *Cell* 1997;89:641-53.

King IF, Francis NJ, Kingston RE. Native and recombinant polycomb group complexes establish a selective block to template accessibility to repress transcription in vitro. *Mol Cell Biol* 2002;22:7919-28.

Kipling D, Davis T, Ostler EL, Faragher RG. What can progeroid syndromes tell us about human aging? *Science* 2004;305:1426-31.

Klein-Szanto AJ, Ruggeri B, Bianchi A, Conti CJ. Cellular and molecular changes during mouse skin tumor progression. *Recent Results Cancer Res* 1993;128:193-204.

Klevecz RR, Shymko RM, Blumenfeld D, Braly PS. Circadian gating of S phase in human ovarian cancer. *Cancer Res* 1987;47:6267-71.

Klevecz RR, Braly PS. Circadian and ultradian cytokinetic rhythms of spontaneous human cancer. *Ann N Y Acad Sci* 1991;618:257-76.

Krebs JE, Fry CJ, Samuels ML, Peterson CL. Global role for chromatin remodeling enzymes in mitotic gene expression. *Cell* 2000;102:587-98.

Krogan NJ, Dover J, et al. The Paf1 complex is required for histone H3 methylation by COMPASS and Dot1p: linking transcriptional elongation to histone methylation. *Mol Cell* 2003;11:721-9.

Krogan NJ, Keogh MC, et al. A Snf2 family ATPase complex required for recruitment of the histone H2A variant Htz1. *Mol Cell* 2003;12:1565-76.

Kume K, Zylka MJ, et al. mCRY1 and mCRY2 are essential components of the negative limb of the circadian clock feedback loop. *Cell* 1999;98:193-205.

Kyng KJ, May A, Kolvraa S, Bohr VA. Gene expression profiling in Werner syndrome closely resembles that of normal aging. *Proc Natl Acad Sci U S A* 2003;100:12259-64.

Lachner M, O'Carroll D, Rea S, Mechtler K, Jenuwein T. Methylation of histone H3 lysine 9 creates a binding site for HP1 proteins. *Nature 2001*;410:116-20.

Lahav G, Rosenfeld N, et al. Dynamics of the p53-Mdm2 feedback loop in individual cells. *Nat Genet* 2004;36:147-50.

Lavigne M, Francis NJ, King IF, Kingston RE. Propagation of silencing; recruitment and repression of naive chromatin in trans by polycomb repressed chromatin. *Mol Cell* 2004;13:415-25.

Lavoie C, Mercier JF, et al. Beta 1/beta 2-adrenergic receptor heterodimerization regulates beta 2-adrenergic receptor internalization and ERK signaling efficacy. *J Biol Chem* 2002;277:35402-10.

Lebel M, Spillare EA, Harris CC, Leder P. The Werner syndrome gene product co-purifies with the DNA replication complex and interacts with PCNA and topoisomerase I. *J Biol Chem* 1999;274:37795-9.

Lee WH, Morton RA, et al. Cytidine methylation of regulatory sequences near the pi-class glutathione S-transferase gene accompanies human prostatic carcinogenesis. *Proc Natl Acad Sci U S A* 1994;91:11733-7.

Lehner AF, Burzynski SR, Hendry LB. 3-Phenylacetylamino-2,6-piperidinedione, a naturally-occurring peptide analogue with apparent antineoplastic activity, may bind to DNA. *Drugs Exp Clin Res* 1986;12 Suppl 1:57-72.

Leone G, Teofili L, Voso MT, Lubbert M. DNA methylation and demethylating drugs in myelodysplastic syndromes and secondary leukemias. *Haematologica* 2002;87:1324-41.

Lessard J, Sauvageau G. Polycomb group genes as epigenetic regulators of normal and leukemic hemopoiesis. *Exp Hematol* 2003;31:567-85.

Li DM, Sun H. TEP1, encoded by a candidate tumor suppressor locus, is a novel protein tyrosine phosphatase regulated by transforming growth factor beta. *Cancer Res* 1997;57:2124-9.

Li E. Chromatin modification and epigenetic reprogramming in mammalian development. *Nat Rev Genet* 2002;3:662-73.

Li H, Zhao L, Yang Z, Funder JW, Liu JP. Telomerase is controlled by protein kinase Calpha in human breast cancer cells. *J Biol Chem* 1998;273:33436-42.

Li Q, Verma IM. NF-kappaB regulation in the immune system. *Nat Rev Immunol* 2002;2:725-34.

Liau MC, Burzynski SR. Altered methylation complex isozymes as selective targets for cancer chemotherapy. *Drugs Exp Clin Res* 1986;12 Suppl 1:77-86.

Liau, MC, Liau, CP, Burzynski SR. Potentiation of induced terminal differentiation by phenylacetic acid and related chemicals. *Internat J Expet Clin Chemother* 1992;5:9-17.

Lin JM, Kilman VL, et al. A role for casein kinase 2alpha in the Drosophila circadian clock. *Nature* 2002;420:816-20.

Lin SY, Elledge SJ. Multiple tumor suppressor pathways negatively regulate telomerase. *Cell* 2003;113:881-9.

Litt MD, Simpson M, Recillas-Targa F, Prioleau MN, Felsenfeld G. Transitions in histone acetylation reveal boundaries of three separately regulated neighboring loci. *Embo J* 2001;20:2224-35.

Liu L, Samid D. Mutant p53 as a target of phenylacetate in human glioblastoma [abstract 2580]. Proceedings of the 86th Annual Meeting of the American Association for Cancer Research, Toronto, Canada; *AACR* 1995;433.

Lombard DB, Beard C, et al. Mutations in the WRN gene in mice accelerate mortality in a p53-null background. *Mol Cell Biol* 2000;20:3286-91.

Longo VD, Finch CE. Evolutionary medicine: from dwarf model systems to healthy centenarians? *Science* 2003;299:1342-6.

Lundblad V, Szostak JW. A mutant with a defect in telomere elongation leads to senescence in yeast. *Cell* 1989;57:633-43.

Maher ER, Reik W. Beckwith-Wiedemann syndrome: imprinting in clusters revisited. *J Clin Invest* 2000;105:247-52.

Malumbres M, Barbacid M. RAS oncogenes: the first 30 years. *Nat Rev Cancer* 2003;3:459-65.

Marciniak RA, Lombard DB, Johnson FB, Guarente L. Nucleolar localization of the Werner syndrome protein in human cells. *Proc Natl Acad Sci U S A* 1998;95:6887-92.

Martinek S, Inonog S, Manoukian AS, Young MW. A role for the segment polarity gene shaggy/GSK-3 in the Drosophila circadian clock. *Cell* 2001;105:769-79.

Matsuo T, Yamaguchi S, et al. Control mechanism of the circadian clock for timing of cell division in vivo. *Science* 2003;302:255-9.

Mayo LD, Donner DB. A phosphatidylinositol 3-kinase/Akt pathway promotes translocation of Mdm2 from the cytoplasm to the nucleus. *Proc Natl Acad Sci U S A* 2001;98:11598-603.

Meneghini MD, Wu M, Madhani HD. Conserved histone variant H2A.Z protects euchromatin from the ectopic spread of silent heterochromatin. *Cell* 2003;112:725-36.

Mizuguchi G, Shen X, et al. ATP driven exchange of histone H2AZ variant catalyzed by SWR1 chromatin remodeling complex. *Science* 2004;303:343-8.

Monk NA. Oscillatory expression of Hes1, p53, and NF-kappaB driven by transcriptional time delays. *Curr Biol* 2003;13:1409-13.

Muldoon TG, Copland JA, Lehner AF, Hendry LB. Inhibition of spontaneous mouse mammary tumour development by antineoplaston A10. *Drugs Exp Clin Res* 1987;13 Suppl 1:83-8.

Muldoon T, Copland J, Hendry L. Antineoplaston A10 activity on carcinogen-induced rat mammary tumors. *Int J Tissue Reactions* 1990;12 (suppl):51-5.

Murphy CT, McCarroll SA, et al. Genes that act downstream of DAF-16 to influence the lifespan of Caenorhabditis elegans. *Nature 2003*;424:277-83.

Nakayama H, Nakayama K, et al. Isolation and genetic characterization of a thymineless death-resistant mutant of Escherichia coli K12: identification of a new mutation (recQ1) that blocks the RecF recombination pathway. *Mol Gen Genet* 1984;195:474-80.

Nakayama M, Wada M, et al. Hypomethylation status of CpG sites at the promoter region and overexpression of the human MDR1 gene in acute myeloid leukemias. *Blood* 1998;92:4296-307.

Nelson DE, Ihekwaba AE, et al. Oscillations in NF-kappaB signaling control the dynamics of gene expression. *Science* 2004;306:704-8.

Nemoto K, Kondo Y, et al. Modulation of telomerase activity by zinc in human prostatic and renal cancer cells. *Biochem Pharmacol* 2000;59:401-5.

Ng HH, Zhang Y, et al. MBD2 is a transcriptional repressor belonging to the MeCP1 histone deacetylase complex. *Nat Genet* 1999;23:58-61.

Noma K, Allis CD, Grewal SI. Transitions in distinct histone H3 methylation patterns at the heterochromatin domain boundaries. *Science* 2001;293:1150-5.

Nugent CI, Lundblad V. The telomerase reverse transcriptase: components and regulation. *Genes Dev* 1998;12:1073-85.

Ordway JM, Curran T. Methylation matters: modeling a manageable genome. *Cell Growth Differ* 2002;13:149-62.

Oshimo Y, Nakayama H, et al. Promoter methylation of cyclin D2 gene in gastric carcinoma. *Int J Oncol* 2003;23:1663-70.

Owen-Hughes T, Bruno M. Molecular biology. Breaking the silence. *Science* 2004;303:324-5.

Pelengaris S, Khan M, Evan G. c-MYC: more than just a matter of life and death. *Nat Rev Cancer* 2002;2:764-76.

Perez-Losada J, Balmain A. Stem-cell hierarchy in skin cancer. *Nat Rev Cancer* 2003;3:434-43.

Perricone NV, Giampapa VC. Skin and aging; the cosmeceutical approach. In: Giampapa VC (Ed), The Basic Principles and Practice of Anti-Aging Medicine & Age Management for the Aesthetic Surgeon and Physician. *Newark: Giampapa* 2003:125-146.

Phelan ML, Sif S, Narlikar GJ, Kingston RE. Reconstitution of a core chromatin remodeling complex from SWI/SNF subunits. *Mol Cell* 1999;3:247-53.

Porter IM, Khoudoli GA, Swedlow JR. Chromosome condensation: DNA compaction in real time. *Curr Biol* 2004;14:R554-6.

Preitner N, Damiola F, et al. The orphan nuclear receptor REV-ERBalpha controls circadian transcription within the positive limb of the mammalian circadian oscillator. *Cell* 2002;110:251-60.

Prokhortchouk A, Hendrich B, et al. The p120 catenin partner Kaiso is a DNA methylation-dependent transcriptional repressor. *Genes Dev* 2001;15:1613-8.

Prokhortchouk E, Hendrich B. Methyl-CpG binding proteins and cancer: are MeCpGs more important than MBDs? *Oncogene* 2002;21:5394-9.

Qu GZ, Grundy PE, Narayan A, Ehrlich M. Frequent hypomethylation in Wilms tumors of pericentromeric DNA in chromosomes 1 and 16. *Cancer Genet Cytogenet* 1999;109:34-39.

Rea S, Eisenhaber F, et al. Regulation of chromatin structure by site-specific histone H3 methyltransferases. *Nature* 2000;406:593-9.

Ren Q, Gorovsky MA. Histone H2A.Z acetylation modulates an essential charge patch. *Mol Cell* 2001;7:1329-35.

Reppert SM, Weaver DR. Coordination of circadian timing in mammals. *Nature* 2002;418:935-41.

Ringrose L, Paro R. Epigenetic regulation of cellular memory by the Polycomb and Trithorax group proteins. *Annu Rev Genet* 2004;38:413-43.

Roberts CW, Orkin SH. The SWI/SNF complex--chromatin and cancer. *Nat Rev Cancer* 2004;4:133-42.

Robertson KD. DNA methylation and chromatin - unraveling the tangled web. *Oncogene* 2002;21:5361-79.

Rohme D. Evidence for a relationship between longevity of mammalian species and life spans of normal fibroblasts in vitro and erythrocytes in vivo. *Proc Natl Acad Sci U S A* 1981;78:5009-13.

Rorke LB, Packer R, Biegel J. Central nervous system atypical teratoid/rhabdoid tumors of infancy and childhood. *J Neurooncol* 1995;24:21-8.

Roth SY, Denu JM, Allis CD. Histone acetyltransferases. *Annu Rev Biochem* 2001;70:81-120.

Rutter J, Reick M, McKnight SL. Metabolism and the control of circadian rhythms. *Annu Rev Biochem* 2002;71:307-31.

Saez E, Rutberg SE, et al. c-fos is required for malignant progression of skin tumors. *Cell* 1995;82:721-32.

Salk D, Bryant E, Au K, Hoehn H, Martin GM. Systematic growth studies, cocultivation, and cell hybridization studies of Werner syndrome cultured skin fibroblasts. *Hum Genet* 1981;58:310-6.

Samid D, Shack S, Sherman LT. Phenylacetate: a novel nontoxic inducer of tumor cell differentiation. *Cancer Res* 1992;52:1988-92.

Sato N, Maitra A, et al. Frequent hypomethylation of multiple genes overexpressed in pancreatic ductal adenocarcinoma. *Cancer Res* 2003;63:4158-66.

Schmitz U, Mueller W, et al. INI1 mutations in meningiomas at a potential hotspot in exon 9. *Br J Cancer* 2001;84:199-201.

Schramke V, Allshire R. Hairpin RNAs and retrotransposon LTRs effect RNAi and chromatin-based gene silencing. *Science* 2003;301:1069-74.

Seitz CS, Lin Q, Deng H, Khavari PA. Alterations in NF-kappaB function in transgenic epithelial tissue demonstrate a growth inhibitory role for NF-kappaB. *Proc Natl Acad Sci U S A* 1998;95:2307-12.

Sephton SE, Sapolsky RM, Kraemer HC, Spiegel D. Diurnal cortisol rhythm as a predictor of breast cancer survival. *J Natl Cancer Inst* 2000;92:994-1000.

Sevenet N, Lellouch-Tubiana A, et al. Spectrum of hSNF5/INI1 somatic mutations in human cancer and genotype-phenotype correlations. *Hum Mol Genet* 1999;8:2359-68.

Shack S, Chen LC, Miller AC, Danesi R, Samid D. Increased susceptibility of ras-transformed cells to phenylacetate is associated with inhibition of p21ras isoprenylation and phenotypic reversion. *Int J Cancer* 1995;63:124-9.

Shao Z, Raible F, et al. Stabilization of chromatin structure by PRC1, a Polycomb complex. *Cell* 1999;98:37-46.

Shay JW, Bacchetti S. A survey of telomerase activity in human cancer. *Eur J Cancer* 1997;33:787-91.

Shay JW, Wright WE. Hayflick, his limit, and cellular ageing. *Nat Rev Mol Cell Biol* 2000;1:72-6.

Shearman LP, Zylka MJ, Weaver DR, Kolakowski LF, Jr., Reppert SM. Two period homologs: circadian expression and photic regulation in the suprachiasmatic nuclei. *Neuron* 1997;19:1261-9.

Shi Y, Lan F, et al. Histone demethylation mediated by the nuclear amine oxidase homolog LSD1. *Cell* 2004;119:941-53.

Siegel PM, Massague J. Cytostatic and apoptotic actions of TGF-beta in homeostasis and cancer. *Nat Rev Cancer* 2003;3:807-21.

Simon JA, Tamkun JW. Programming off and on states in chromatin: mechanisms of Polycomb and trithorax group complexes. *Curr Opin Genet Dev* 2002;12:210-8.

Sims RJ, 3rd, Reinberg D. From chromatin to cancer: a new histone lysine methyltransferase enters the mix. *Nat Cell Biol* 2004;6:685-7.

Singer MF, Krek V, McMillan JP, Swergold GD, Thayer RE. LINE-1: a human transposable element. *Gene* 1993;135:183-8.

Sommers JA, Sharma S, et al. p53 modulates RPA-dependent and RPA-independent WRN helicase activity. *Cancer* Res 2005;65:1223-33.

Soultanas P, Wigley DB. Unwinding the 'Gordian knot' of helicase action. *Trends Biochem Sci* 2001;26:47-54.

Spillare EA, Robles AI, et al. p53-mediated apoptosis is attenuated in Werner syndrome cells. *Genes Dev* 1999;13:1355-60.

Springhetti EM, Istomina NE, et al. Role of the M-loop and reactive center loop domains in the folding and bridging of nucleosome arrays by MENT. *J Biol Chem* 2003;278:43384-93.

Stephan FK, Zucker I. Circadian rhythms in drinking behavior and locomotor activity of rats are eliminated by hypothalamic lesions. *Proc Natl Acad Sci U S A* 1972;69:1583-6.

Stokkan KA, Yamazaki S, Tei H, Sakaki Y, Menaker M. Entrainment of the circadian clock in the liver by feeding. *Science* 2001;291:490-3.

Strahl BD, Ohba R, Cook RG, Allis CD. Methylation of histone H3 at lysine 4 is highly conserved and correlates with transcriptionally active nuclei in Tetrahymena. *Proc Natl Acad Sci U S A* 1999;96:14967-72.

Sudarsanam P, Iyer VR, Brown PO, Winston F. Whole-genome expression analysis of snf/swi mutants of Saccharomyces cerevisiae. *Proc Natl Acad Sci U S A* 2000;97:3364-9.

Sun SC, Ganchi PA, Ballard DW, Greene WC. NF-kappa B controls expression of inhibitor I kappa B alpha: evidence for an inducible autoregulatory pathway. *Science* 1993;259:1912-5.

Sun ZS, Albrecht U, et al. RIGUI, a putative mammalian ortholog of the Drosophila period gene. *Cell* 1997;90:1003-11.

Surh YJ. Cancer chemoprevention with dietary phytochemicals. *Nat Rev Cancer* 2003;3:768-80.

Suter CM, Martin DI, Ward RL. Hypomethylation of L1 retrotransposons in colorectal cancer and adjacent normal tissue. *Int J Colorectal Dis* 2004;19:95-101.

Takai D, Jones PA. Comprehensive analysis of CpG islands in human chromosomes 21 and 22. *Proc Natl Acad Sci U S A* 2002;99:3740-5.

Takakura M, Kyo S, et al. Telomerase activation by histone deacetylase inhibitor in normal cells. *Nucleic Acids Res* 2001;29:3006-11.

Takano A, Shimizu K, et al. Cloning and characterization of rat casein kinase 1epsilon. *FEBS Lett* 2000;477:106-12.

Takumi T, Taguchi K, et al. A light-independent oscillatory gene mPer3 in mouse SCN and OVLT. *Embo J* 1998;17:4753-9.

Talwar HS, Griffiths CE, Fisher GJ, Hamilton TA, Voorhees JJ. Reduced type I and type III procollagens in photodamaged adult human skin. J Invest Dermatol 1995;105:285-90.

Tamaru H, Selker EU. A histone H3 methyltransferase controls DNA methylation in Neurospora crassa. *Nature* 2001;414:277-83.

Taylor MD, Gokgoz N, et al. Familial posterior fossa brain tumors of infancy secondary to germline mutation of the hSNF5 gene. *Am J Hum Genet* 2000;66:1403-6.

Tei H, Okamura H, et al. Circadian oscillation of a mammalian homologue of the Drosophila period gene. *Nature* 1997;389:512-6.

Thiru A, Nietlispach D, et al. Structural basis of HP1/PXVXL motif peptide interactions and HP1 localisation to heterochromatin. *Embo J* 2004;23:489-99.

Trouche D, Le Chalony C, Muchardt C, Yaniv M, Kouzarides T. RB and hbrm cooperate to repress the activation functions of E2F1. *Proc Natl Acad Sci U S A* 1997;94:11268-73.

Trush MA, Kensler TW. An overview of the relationship between oxidative stress and chemical carcinogenesis. *Free Radic Biol Med* 1991;10:201-9.

Turker MS. Gene silencing in mammalian cells and the spread of DNA methylation. *Oncogene* 2002;21:5388-93.

Ueda HR, Chen W, et al. A transcription factor response element for gene expression during circadian night. *Nature* 2002;418:534-9.

Ushijima T. Detection and interpretation of altered methylation patterns in cancer cells. *Nat Rev Cancer* 2005;5:223-31.

van Steensel B, Smogorzewska A, de Lange T. TRF2 protects human telomeres from end-to-end fusions. *Cell* 1998;92:401-13.

Velicescu M, Weisenberger DJ, et al. Cell division is required for de novo methylation of CpG islands in bladder cancer cells. *Cancer Res* 2002;62:2378-84.

Versteege I, Medjkane S, Rouillard D, Delattre O. A key role of the hSNF5/INI1 tumour suppressor in the control of the G1-S transition of the cell cycle. *Oncogene* 2002;21:6403-12.

Villa R, Porta CD, Folini M, Daidone MG, Zaffaroni N. Possible regulation of telomerase activity by transcription and alternative splicing of telomerase reverse transcriptase in human melanoma. *J Invest Dermatol* 2001;116:867-73.

Villuendas R, Sanchez-Beato M, et al. Loss of p16/INK4A protein expression in non-Hodgkin's lymphomas is a frequent finding associated with tumor progression. *Am J Pathol* 1998;153:887-97.

Vivanco I, Sawyers CL. The phosphatidylinositol 3-Kinase AKT pathway in human cancer. *Nat Rev Cancer* 2002;2:489-501.

Vousden KH, Lu X. Live or let die: the cell's response to p53. *Nat Rev Cancer* 2002;2:594-604.

Waikel RL, Kawachi Y, Waikel PA, Wang XJ, Roop DR. Deregulated expression of c-Myc depletes epidermal stem cells. *Nat Genet* 2001;28:165-8.

Waldbillig R, Burzynski SR. Mechanism of action, uptake, and gene array studies on the antineoplastic agent phenylacetylglutamine (PG) in human glioma cells U-87. *Neuro-oncol* 2003;5:309.

Waldbillig RJ, Patil S, Burzynski SR. Uptake and intracellular binding of the antineoplastic agents phenylacetic acid and phenylacetylglutamine: effects on epigenetic mechanisms of gene regulation and gene expression. *Report to FDA*. 2004.

Wan YS, Wang ZQ, Shao Y, Voorhees JJ, Fisher GJ. Ultraviolet irradiation activates PI 3-kinase/AKT survival pathway via EGF receptors in human skin in vivo. *Int J Oncol* 2001;18:461-6.

Wang XW, Tseng A, et al. Functional interaction of p53 and BLM DNA helicase in apoptosis. *J Biol Chem* 2001;276:32948-55.

Waterland RA, Jirtle RL. Transposable elements: targets for early nutritional effects on epigenetic gene regulation. *Mol Cell Biol* 2003;23:5293-300.

Webster LT, Siddiqui UA, Lucas SV, Strong JM, Mieyal JJ. Identification of separate acyl-CoA:glycine and acyl-CoA:L-glutamine N-acyltransferase activities in mitochondrial fractions from liver of rhesus monkey and man. *J Biol Chem* 1976;251:3352-8.

Weng LP, Brown JL, Eng C. PTEN coordinates G(1) arrest by down-regulating cyclin D1 via its protein phosphatase activity and up-regulating p27 via its lipid phosphatase activity in a breast cancer model. *Hum Mol Genet* 2001;10:599-604.

Werth VP, Williams KJ, et al. UVB irradiation alters cellular responses to cytokines: role in extracellular matrix gene expression. *J Invest Dermatol* 1997;108:290-4.

Williams JA, Su HS, Bernards A, Field J, Sehgal A. A circadian output in Drosophila mediated by neurofibromatosis-1 and Ras/MAPK. *Science* 2001;293:2251-6.

Winston F, Carlson M. Yeast SNF/SWI transcriptional activators and the SPT/SIN chromatin connection. *Trends Genet* 1992;8:387-91.

Wu KJ, Grandori C, et al. Direct activation of TERT transcription by c-MYC. *Nat Genet* 1999;21:220-4.

Wyllie FS, Jones CJ, et al. Telomerase prevents the accelerated cell ageing of Werner syndrome fibroblasts. *Nat Genet* 2000;24:16-7.

Yasumoto S, Kunimura C, et al. Telomerase activity in normal human epithelial cells. *Oncogene* 1996;13:433-9.

Yeh A, Wei M, et al. Chromosome arm 16q in Wilms tumors: unbalanced chromosomal translocations, loss of heterozygosity, and assessment of the CTCF gene. *Genes Chromosomes Cancer* 2002;35:156-63.

Yoder JA, Walsh CP, Bestor TH. Cytosine methylation and the ecology of intragenomic parasites. *Trends Genet* 1997;13:335-40.

Young MW, Kay SA. Time zones: a comparative genetics of circadian clocks. *Nat Rev Genet* 2001;2:702-15.

Yui J, Chiu CP, Lansdorp PM. Telomerase activity in candidate stem cells from fetal liver and adult bone marrow. *Blood* 1998;91:3255-62.

Zelent A, Petrie K, et al. Molecular target-based treatment of human cancer: summary of the 10th international conference on differentiation therapy. *Cancer Res* 2005;65:1117-23.

Zhang HS, Gavin M, et al. Exit from G1 and S phase of the cell cycle is regulated by repressor complexes containing HDAC-Rb-hSWI/SNF and Rb-hSWI/SNF. *Cell* 2000;101:79-89.

Zhou BP, Liao Y, et al. Cytoplasmic localization of p21Cip1/WAF1 by Akt-induced phosphorylation in HER-2/neu-overexpressing cells. *Nat Cell Biol* 2001;3:245-52.

Zhu Y, Parada LF. The molecular and genetic basis of neurological tumours. *Nat Rev Cancer* 2002;2:616-26.

In: Gene Silencing: New Research
Editor: Grace W. Redberry, pp. 81-119

ISBN 1-59454-832-3

Chapter III

Gene Silencing in Tumors: The Role of the Ras Association Domain Family 1 Gene in Carcinogenesis

Reinhard Dammann*
AG Tumorgenetik der Medizinischen Fakultät,
Martin-Luther-Universität Halle-Wittenberg, Germany

Abstract

Epigenetic silencing of tumor suppressor genes plays an important role in the pathogenesis of human cancer. Especially, hypermethylation of the promoter regions of cancer-related genes is often observed in tumors. Recently, a novel tumor suppressor gene, which was termed Ras association domain family 1 (RASSF1A) gene was characterized by its frequent epigenetic silencing in human tumors. RASSF1 consists of two main variants (RASSF1A and RASSF1C), which are transcribed from distinct CpG island promoters. Aberrant methylation of the RASSF1A promoter region is one of the most frequent epigenetic inactivation events detected in human cancer and leads to silencing of RASSF1A. Silencing of RASSF1A was commonly observed in primary tumors including lung, breast, pancreas, kidney, liver, cervix, nasopharyngeal, prostate, thyroid and other cancers. Moreover, RASSF1A methylation was frequently detected in body fluids including blood, urine, nipple aspirates, sputum and bronchial alveolar lavages. Inactivation of RASSF1A was associated with an advanced tumor stage (e.g. bladder, brain, prostate, gastric tumors) and poor prognosis (e.g. lung, sarcoma and breast cancer). Detection of aberrant RASSF1A methylation may serve as a diagnostic and prognostic marker. The functional analyses of RASSF1A reveal an involvement in apoptotic signaling, microtubule stabilization and mitotic progression. The tumor suppressor RASSF1A may act as a negative Ras effector inhibiting cell growth and

* Correspondence: AG Tumorgenetik der Medizinischen Fakultät, Martin-Luther-Universität Halle-Wittenberg, Magdeburger Straße 2, 06097 Halle/Saale, Germany; Tel: 49-345-557-4537; Fax: 49-345-557-4293; E-mail: reinhard.dammann@medizin.uni-halle.de

inducing cell death. Thus, RASSF1A represents a tumor suppressor gene, which is frequently epigenetically silenced in the pathogenesis of human cancers.

Abbreviations

RASSF	Ras-association domain family;
NORE	novel Ras effector;
LOH	loss of heterozygosity;
RA domain	RalGDS/AF6 Ras-association domain;
aa	amino acid;
5-aza-CdR	5-aza-2'-deoxycytidine;
SV40	simian virus 40;
EBV	Epstein-Barr virus;
HPV	human papilloma virus,
ATM	ataxia telangiectasia mutated;
DAG	diacylglycerol;
C1	protein kinase C conserved region 1;
SCC	squamous cell carcinoma;
AC	adenocarcinoma;
SCLC	small cell lung cancer;
NSCLC	non-small cell lung cancer.

Introduction

Epigenetic silencing of tumor suppressor gene promoters plays a fundamental role in the etiology of cancer [85]. Inactivation of cancer-related genes is mediated through remodeling of chromatin structure, which is correlated to changes in DNA methylation and histone modifications. Gene promoters are often located in CpG-rich DNA regions, so called CpG islands [17]. Active promoters are associated with unmethylated CpG islands and open chromatin structure for transcription regulators (Figure 1). Whereas inactive promoters are in a repressed chromatin structure and the CpGs are hypermethylated (Figure 1). Acetylation of lysines at histones H3 leads to active chromatin and deacetylation is correlated with a repressed chromatin structure [81]. Methylation of histone lysine 9 residue of histone H3 is associated with promoters of inactive genes (Figure 1). Methyl-CpG binding proteins (MBPs) provide a link between methylated DNA and hypoacetylated histones by recruiting histone deacetylases [86, 136]. DNA methyltransferases, which are responsible for the maintenance of methylated DNA and necessary for the establishment of newly methylated promoters, were shown to be associated with proteins including Rb, E2F1, histone deacetylases and a transcriptional repressor [55, 148, 151]. It has been proposed the DNA methylation may be directed by alterations in the chromatin structure [16]. Several studies indicate an altered DNA methylation pattern when components of the chromatin-remodeling system, such as SNF2 like factors were mutated [37, 61, 80]. In *Neurospora crassa* and in *Arabidopsis thaliana,* it has been shown that a repressive chromatin modification, like histone H3 lysine 9 (H3-K9) methylation can direct DNA methylation [78, 170]. In vertebrate, it has been reported that Suv39h-mediated H3-K9 methylation directs DNA methylation to major satellite repeats at pericentric heterochromatin [112]. It has been shown that methylated CpG sites attract methyl CpG binding proteins (MBPs) that interact with the co-repressor complex Sin3 including histone deacetylases [86, 136, 138]. Other studies indicate that DNA methyltransferase and MBP interact with histone methlytransferase [56, 57]. In cancer cells, aberrant DNA methylation may maintain the repressed epigenetic state. This may lead to silencing of tumor suppressor genes, which is irreversible since the presence of an active DNA demethylase in mammalian cells is not evident. It has been shown that the heterochromatin protein 1 (Hp1) isoforms bind to methylated H3-K9 residues [11, 107, 135]. Recently, a link between the Suv39h-Hp1 histone methylation system and the DNA methyltransferase 3b in mammals was demonstrated [112]. Thus Hp1 and MBD may be responsible to lock and maintain the repressed chromatin together with the DNA methyltransferases in silenced cells (Figure 1).

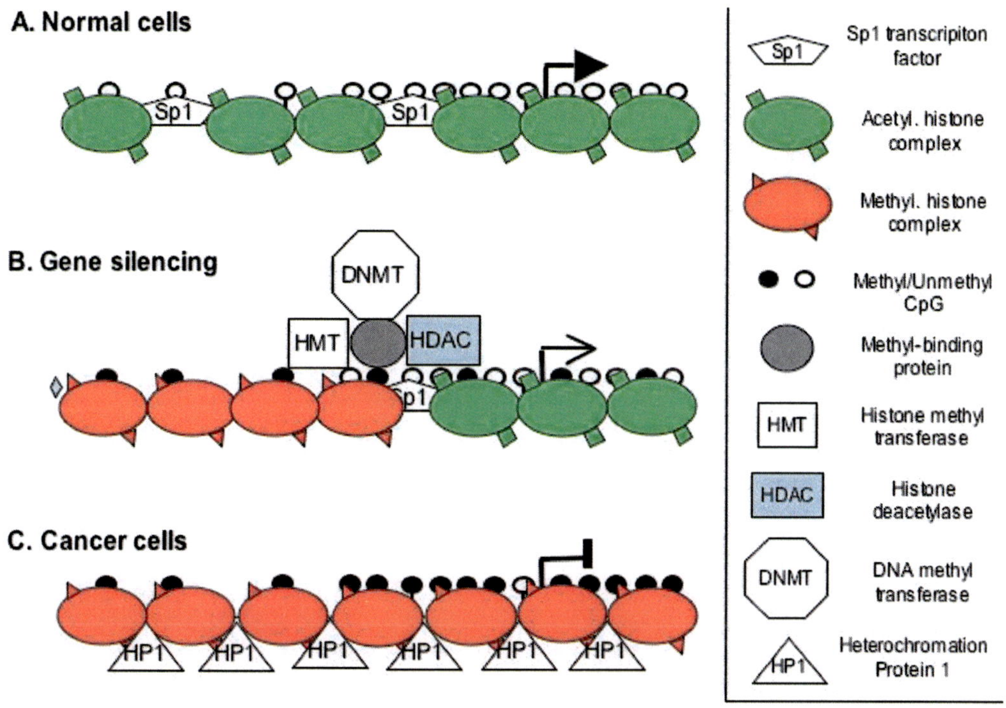

Figure 1. Model of the progressive epigenetic silencing in cancer. In normal epithelial cells, the promoter is transcriptionally active and histones are acetylated and CpGs island is unmethylated (white dots). The open chromatin structure allows binding of the transcription factor Sp1. In senescent cells, silencing is associated with histone deacetylation and methylatio, which is accomplished by histone deacteylase (HDAC) and histone methyltransferase (HMT), respectively. The repressed chromatin structured triggers the *de novo* methylation of CpGs (black dots) by DNA methyl transferase (DNMT). In cancer cells, the inactive state is locked and maintained by Methyl-CpG binding domain proteins (MBD) and heterochromatin protein 1 (Hp1).

In a model of aging and cancer it has been postulated that an age-related CpG island methylation of tumor suppressor genes leads to a hyperproliferative epithelium, but further hypermethylation needs to occur for tumor formation [77]. It has been suggested that a dramatic stop in transcription is the critical precursor in cancer, which is followed by *de novo* methylation and ends with a complete cessation of gene expression [27, 187]. Recent studies have demonstrated that the CpG islands of the Rb, p16, VHL, APC, MLH1 and BRCA1 genes are frequently methylated in a variety of human cancers, but are largely methylation-free in the corresponding normal tissues [4, 85]. It is remarkable that as many, if not more, tumor suppressor genes are silenced by promoter hypermethylation as they are by coding-region mutations. Therefore, epigenetic silencing of tumor suppressor genes plays a key role in human carcinogenesis. Recently, the novel tumor suppressor RASSF1A was identified, which is frequently inactivated in a variety of human cancers. Silencing of RASSF1A is one of the most common epigenetic events, which occurs in the pathogenesis of human tumors.

Identification and Characterization of *RASSF1*

Loss of the short arm of chromosome 3 is the earliest and most common alteration, which occurs in the pathogenesis of lung cancer. Several distinct regions are lost, including 3p12, 3p14, 3p21 and 3p24-25 [99]. In these segments, the van Hippel-Lindau disease (*VHL*) gene at 3p25 [87], the gene *FHIT* at 3p14.2 [167], and the *DUTT1/ROBO1* gene at 3p12 [206] have been identified. At segment 3p21.3 heterozygous and homozygous deletions have been described in several cancer cell lines and in primary lung tumors [95, 99, 172, 197, 198, 209]. Allelic loss at 3p21.3 is not limited to lung cancer indicating that this segment may encode a general tumor suppressor. Other tumors with 3p21 involvement include head and neck cancer, renal cell carcinoma, bladder cancer, female genital tract tumors and breast cancer [99]. The region of minimum homozygous deletion at 3p21.3 was narrowed to a fragment of 120 kb using several cancer cell lines [158].

Recently, we and others have cloned the RASSF1 gene from the common homozygous deletion area at 3p21.3 [30, 113]. RASSF1 was isolated through the interaction with the human DNA excision repair protein XPA in a yeast two-hybrid screen [30]. The 1.7 kb cDNA matched the sequences of the minimum homozygous deletion region of 120 kb [158] that may contain the tumor suppressor gene at 3p21.3. The C-terminus of RASSF1 shows high homology (ca. 55% identity) to the murine Ras-effector protein Nore1 [190] and encodes a Ras-association domain (Figure 2). Therefore, the gene has been termed Ras-association domain family 1 gene [30]. Homology searches and additional cDNA screenings revealed the presence of seven alternatively spliced transcripts: RASSF1A, RASSF1B, RASSF1C, RASSF1D, RASSF1E, RASSF1F and RASSF1G [30, 34]. The two major forms RASSF1A and RASSF1C are transcribed from two distinct CpG island promoters, which are approximately 3.5 kb apart (Figure 2). Both transcripts have four common exons (3 to 6) at their 3' end (Figure 2). These exons encode the RA domain (R194 to S288) [141]. RASSF1A has two 5' exons, designated 1α and 2αβ. The cDNA is 1.9 kb long and contains an ORF of 340 amino acids (aas) with a calculated MW of 38.8 kDa. The N-terminus (H52 to C101) of RASSF1A has high homology to a cysteine-rich diacylglycerol/phorbol ester-binding (DAG) domain, also known as the protein kinase C conserved region (C1), which contains a central C1 zinc finger [137]. RASSF1A is expressed in all normal tissues tested by Northern blot analysis, but was missing in several cancer cell lines [30]. Transcript RASSF1C is 1.7 kb long and transcription initiates in exon 2γ located at the CpG island C (Figure 2). The cDNA encodes a 270 aas protein with an MW of 31.2 kDa. The protein sequence translated from the first exon 2γ has no significant similarity to any known protein. RASSF1C is transcribed in all normal tissue and cancer cells tested [30]. The aa sequence W125 to K138 (WETPDLSQAEIEQK) of RASSF1A matches a putative ATM kinase phosphorylation consensus motif and in a peptide with this sequence serine is effectively phosphorylated *in vitro* [98]. Recent data show that RASSF1A interacts with the microtubule and stabilizes the cytoskeleton and centrosome [121]. The microtubule association and stabilization domain (MAS) was located at the C-terminus (194 to 288) of RASSF1A (Figure 2).

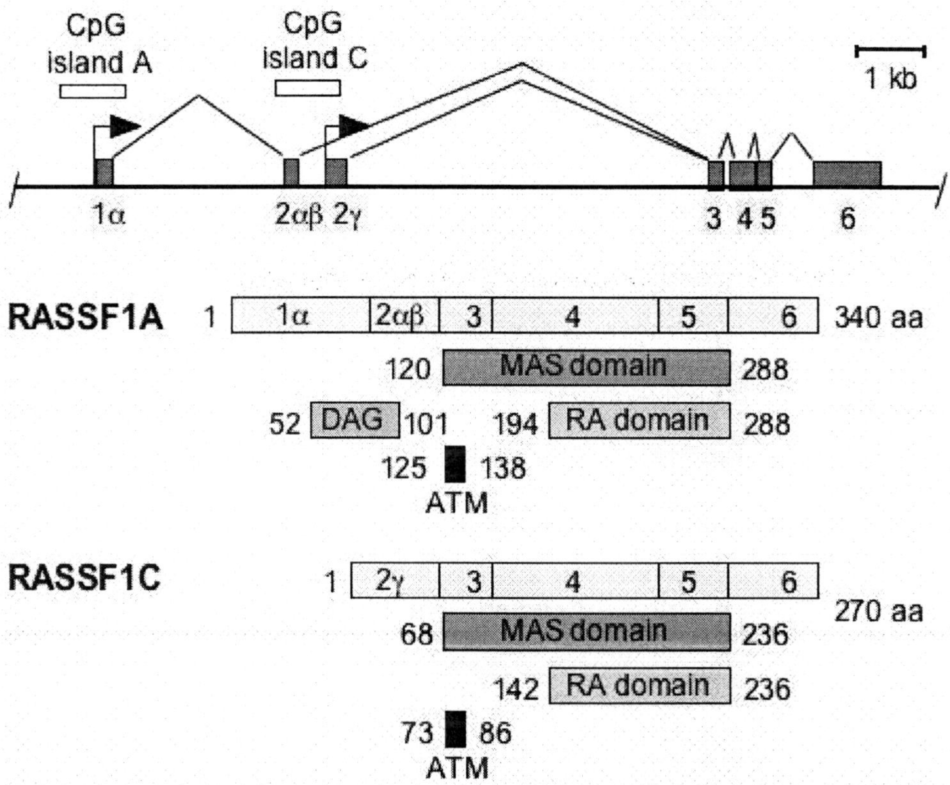

Figure 2. Map of the RASSF1 gene. The two promoters of RASSF1 (arrows) are located in CpG islands (open squares). Two major isoforms (RASSF1A and RASSF1C) are made by alternative promoter usage and splicing of the exons (black boxes). The cDNA of RASSF1A is 1.9 kb long and contains an ORF of 340 amino acids (aas) with a calculated MW of 38.8 kDa. Transcript RASSF1C is 1.7 kb long and initiates in exon 2γ located at the CpG island C. The cDNA encodes a 270 aas protein with a MW of 31.2 kDa. The protein domains are indicated as: DAG, diacylglycerol/phorbol ester binding domain; RA, RalGDS/AF6 Ras-association domain; and ATM, putative ATM phosphorylation site consensus sequence; MAS, microtubule association and stabilization domain.

The expression of RASSF1A and RASSF1C have been analyzed in a variety cell lines and tissues. In all normal cells both forms, RASSF1A and RASSF1C, are highly expressed [30]. RASSF1A expression is missing in several cancer cell lines including non-small cell lung cancer (NSCLC), small cell lung cancer (SCLC), breast carcinoma, nasopharyngeal carcinoma, renal cell carcinoma, and thyroid carcinoma [18, 30-32, 42, 122, 154]. In contrast, RASSF1C was expressed in all analyzed samples without homozygous deletion of 3p21.3 [18, 30-32, 154]. Loss of expression was correlated with hypermethylation of the CpG island of the RASSF1A promoter [1, 30]. In addition, reversion of the epigenetic silencing by treatment with a DNA methylation inhibitor, 5-aza-2'-deoxycytidine (5-aza-CdR), leads to re-expression of RASSF1A in various cancer cell lines [8, 18, 19, 30, 32, 122, 154].

Methylation of RASSF1A in Human Tumors

Epigenetic silencing of tumor suppressor genes by aberrant methylation is a fundamental inactivation mechanism of cancer-related genes in the pathogenesis of human cancer [66, 85]. Promoter hypermethylation of RASSF1A was frequently detected in primary human tumors and this methylation was correlated with several additional observations (Table 1). In bladder cancer, high frequency of RASSF1A methylation was observed and was correlated with advanced tumor stage and poor prognosis [21, 45, 108, 126]. In brain cancer, RASSF1A methylation was often detected in neuroblastoma, astrocytoma, glioblastoma and medulloblastoma, however it was less frequent found in benign tumors [8-10, 60, 68, 73, 118, 123, 145, 212, 215]. In benign ganglioneuroma, RASSF1A methylation was not observed [212]. Aberrant methylation of RASSF1A was rarely detected in gallbladder carcinoma and chronic cholecystitis, but was frequently observed in bile duct carcinoma and ampullary carcinoma [169, 185]. Methylation of RASSF1A was frequently found in breast cancer and in serum of breast cancer patients [1, 18, 23, 32, 49, 70, 102, 111, 131, 133]. In cervical cancer, RASSF1A promoter methylation was found in certain tumor types (Table 1) [1, 28, 92, 105, 216]. Kuzmin et al. (2003) have reported an inverse correlation between methylation of RASSF1A and human papilloma virus infection. In cholangiocarcinoma, 65 to 69% of RASSF1A promoter hypermethylation were observed [201, 211]. In colorectal carcinoma, RASSF1A methylation was less frequently found [110, 189, 194, 208, 214]. Interestingly, RASSF1A methylation occurs significantly in colorectal carcinoma with K-ras wild type [152, 188]. In 52% of esophageal squamous cell carcinoma, RASSF1A methylation was reported and correlated with an advanced tumor stage [103]. In gastric cancer, RASSF1A hypermethylation was more frequently found in the advanced tumor stage and in EBV positive carcinoma (Table 1); however, RASSF1A methylation was rarely detected in non-carcinoma tumor samples [19, 74, 88-90, 171]. In head and neck cancer, RASSF1A methylation frequency is less than 20% [40, 65, 69, 125]. Dong el al. 2003 have reported an inverse correlation between RASSF1A methylation and HPV infection (Table 1). In hepatocellular carcinoma, intensive RASSF1A methylation was detected and methylation was also found in adjacent non-cancerous tissue, cirrhosis and hepatitis [109, 155, 216, 217]. In Hodgkins' lymphoma, 65% of RASSF1A methylation was reported [134]. Several reports have investigated the methylation status of RASSF1A in renal cell carcinoma and kidney tumors [13, 42, 43, 62, 132, 173, 214]. In papillary renal cell carcinoma, RASSF1A was frequently inactivated (Table 1). Intensive methylation (>70%) of RASSF1A was reported in small cell lung cancer (SCLC). In non small cell lung cancer (NSCLC), RASSF1A hypermethylation is common and correlated with impaired prognosis and with starting smoking under age 18 [1, 18, 30, 31, 47, 70, 76, 96, 97, 115, 124, 128, 145, 174, 179, 181, 183, 196, 210, 218]. In malignant mesothelioma, 32% of RASSF1A inactivation was found and correlated with the presence of SV40 [180, 182]. In melanoma, frequent RASSF1A methylation was reported in several studies [58, 72, 147, 168]. Reifenberger et al. (2004) observed that tumors with RASSF1A methylation additionally carried BRAF and NRAS mutations, suggesting a synergistic effect of these aberrations on melanoma growth. In less than 30% of multiple myeloma cases, RASSF1A methylation was found [59, 139, 157]. In nasopharyngeal carcinoma, aberrant RASSF1A methylation was frequently (> 50%) observed [22, 106, 122, 178, 202, 203] and occurred in 39% of EBV associated nasopharyngeal carcinoma (Table 1). In 40% of osteosarcoma, hypermethylation of RASSF1A occurred [117]. In ovarian cancer frequent RASSF1A methylation was demonstrated in several studies [1, 36, 38, 146, 214] and also detected in granulosa cell tumors [38]. In endocrine tumors of the pancreas, the frequency of RASSF1A inactivation was higher compared to pancreatic adenocarcinoma (83% versus 64%) [33, 75]. We have reported an inverse correlation between RASSF1A silencing and K-ras mutation in pancreatic cancer. In pediatric tumors, RASSF1A methylation was found in Wilms' tumor, medulloblastoma, retinoblastoma, rhabdomyosarcoma, neuroblastoma, hepatoblastoma, leukemia, pancreatoblastoma, adrenocortical carcinoma and lymphoma [46, 64, 194, 200]. In hereditary pheochromocytoma, RASSF1A methylation was more common compared to the sporadic tumors [8, 35]. In prostate cancer, high frequency of RASSF1A methylation was reported in several studies and correlated with an advanced Gleason score (Table 1) [52, 82, 91, 104, 120, 127, 162, 204, 205]. However, methylation was also observed in several non-carcinoma specimen from matched tumor-normal samples and in benign prostatic hyperplasia. RASSF1A methylation was detected in different entities of soft tissue sarcoma, including leiomyosarcoma [156]. In testicular germ cell tumors, most studies have detected that RASSF1A methylation occurs frequently [71, 93, 100, 101]. Interestingly, RASSF1A methylation occurred more often in cisplatin-resistant tumors (Table 1). In thyroid cancer, RASSF1A hypermethylation was frequently detected in the more aggressive carcinomas [154, 207]. In papillary thyroid cancer, an inverse correlation between RASSF1A methylation and BRAF mutation was reported (Table 1).

In general, RASSF1A methylation frequency is higher in cancer cell lines compared to the primary tumors. These additional changes could be attributed to *de novo* methylation, which occurs when cells are kept in culture [5, 84, 163]. In cell lines with an inactivated RASSF1A gene treatment with 5-aza-CdR reactivated the expression of RASSF1A (Table 1). In principal, methylation of RASSF1A is rarely detected in normal tissue, however methylation was also found in some non-cancerogenous tissue specimen. For instance RASSF1 methylation was detected in adjacent normal tissue of prostate, kidney, thyroid and liver cancer patients (Table 1). This methylation may represent infiltration of tumor cells into normal tissue or a field defect leading to carcinogenesis.

Table 1. RASSF1A methylation in primary human tumors

Tumor entity	Methylation	Ref.	Remarks
Bladder cancer	60% (33/55)	[108]	Inactivation of RASSF1A was correlated with advanced tumor stage
	35% (34/98)	[126]	RASSF1A methylation correlated with parameters of poor prognosis
	48% (19/40)	[21]	RASSF1A methylation was more frequent in cases with LOH at 3p21.3 (73%) compared to cases without LOH (13%; p=0.007). RASSF1A methylation was found in 50% (7/14) of urine samples; no false positives.
Bladder cancer	51% (23/45)	[45]	RASSF1A hypermethylation was found in all pathological grades and stages of bladder cancer and in patients of all ages; in 87% (39/45) of the cases hypermethylation of RASSF1A, APC or $p14^{ARF}$ could be detected in urine of the bladder cancer patients.
Biliary tract cancer	0% (0/50)	[169]	Gallbladder carcinoma; No methylation of RASSF1A was detected in chronic cholecystitis
	11% (1/9)	[185]	Gallbladder carcinoma
	30% (7/23)	[185]	Bile duct carcinoma
	40% (2/5)	[185]	Ampullary carcinoma
Brain cancer	55% (37/67)	[8]	Neuroblastoma; RASSF1A was reexpressed after treatment with 5-aza- CdR in neuroblastoma cell lines
	54% (25/46)	[73]	Glioma
	100% (5/5)	[73]	Medulloblastoma
	10% (1/10)	[73]	Schwannoma (benign)
	[illegible]	[73]	Meningioma (benign)
	79% (27/34)	[123]	Epigenetic inactivation by biallelic hypermethylation represents the primary mechanism of RASSF1A inactivation in medulloblastoma.
	93% (41/44)	[118]	Medulloblastoma; in 57% of the cases a total methylation was detected; methylation in all histopathological and clinical disease subtypes100% (11/11) in medulloblastoma cell lines and RASSF1A was reexpressed after treatment with 5-aza-CdR
	57% (36/63)	[68]	Glioma; RASSF1A methylation increased with tumor grade (40% grade II, 53% grade III, 63% grade IV); no association between RASSF1A and BLU methylation; 100% (7/7) in glioma cell lines; reexpression of RASSF1A suppressed the growth of glioma cell line.
	57% (12/21)	[10]	Glioblastoma; a tendency for a longer time to progression for patients with methylated RASSF1A promoter was observed; 50% (13/26) in serum samples
	57% (16/28)	[145]	Glioblastoma; 50% in serum; high correlation between methylation in tumor and serum was observed (Spearman test p=0.0001)
	55%	[9]	Neuroblastoma; RASSF1A promoter hypermethylataion was more frequent in neuroblastomas with SLIT2 promoter methylation (p= 0.32); inverse relationship between SLIT2 and RASSF1A promoter hypermethylation in Wilms tumor (p= 0.09); no associations with clinicopathological features
	70% (37/53)	[215]	Astrocytoma; RASSF1C was completely unmethylated
	32% (13/14)	[60]	Glioblastoma

Table 1. RASSF1A methylation in primary human tumors (Contnued)

Tumor entity	Methylation	Ref.	Remarks
Brain cancer	70% (39/56)	[212]	Neuroblastoma; RASSF1A methylation was not found in benign ganglioneuromas; RASSF1A methylation was associated with age(>1; p<0.01), high-risk disease (p<0.016) and poor survival (p<0.001)
Breast cancer	62% (28/45)	[32]	RASSF1A was reexpressed after treatment with 5-aza-CdR in breast cancer cell lines
	9% (4/44)	[1]	
	49% (19/39)	[18]	
	56% (20/36)	[111]	RASSF1A methylation was detected in epithelial hyperplasia, but not in normal tissue
	58% (54/93)	[23]	
	65% (11/17)	[70]	Invasive breast cancer
	42% (5/12)	[70]	Ductal carcinoma in-situ
	62% (8/13)	[49]	Lobular carcinoma in-situ
	84% (16/19)	[49]	Invasive lobular cancer
	70% (19/27)	[49]	Invasive breast cancer
	75%	[49]	Ductal carcinoma in-situ
	23% (n=26)	[133]	Serum DNA from primary breast cancer patients: patients with methylated RASSF1A and/or APC serum DNA was strongly associated with poor outcome, with a relative risk for death of 5.7 (p<0.001); 80% (n=10) in serum DNA from recurrent breast cancer patients).
	56% (14/25)	[131]	Primary breast cancer, additional analyses of metastases : 78% (7/9) bone; 67% (4/6) brain; 100% (10/10) lung; higher prevalence of methylation in lymph node metastasis than in primary tumors
	62%	[102]	Hypermethylation in nipple aspirates was detected in matched breast tumor cases
	65% (22/34)	[44]	Methylation of RASSF1A was only detected in serum DNA of breast cancer patients with methylated tumor
Cervix cancer	0% (0/22)	[1]	
	30% (10/33)	[216]	Squamous cell carcinoma; 41 % hypermethylation among the group of SCC with 3p21 allelic loss whereas only 21% of SCC with retention of 3p21 demonstrated RASSF1A hypermethylation
	12% (2/17)	[216]	Adenocarcinoma ; no correlation between RASSF1A hypermethylation and age of patient, HPV genotype, tumor grade or stage was observed
	10% (4/42)	[105]	Squamous cell carcinoma
	21% (4/19)	[105]	Adenosquamous carcinoma
	24% (8/34)	[105]	Adenocarcinoma; significant reverse correlation between inactivation of RASSF1A and the presence of high-risk HPV was observed in cervical tumors and cell lines (p<0.04).
	0% (0/31)	[28]	Squamous cell carcinoma
	12% (10/82)	[92]	Uterine cervical cancer

Table 1. RASSF1A methylation in primary human tumors (Contnued)

Tumor entity	Methylation	Ref.	Remarks
Cholangio-carcinoma	69% (9/13)	[201]	Expression of RASSF1A in nine cases with promoter methylation indicated reduced expression compared to normal livers.
	63% (47/72)	[211]	Methylation of RASSF1A was more common in extrahepatic than intrahepatic cholangiocarcinoma (p=0.003)
Colorectal cancer	12% (3/26)	[214]	
	20% (45/222)	[188]	RASSF1A methylation occurs predominatly in K-ras wild type colorectal carcinomas (P=0.023)
	45% (13/29)	[194]	RASSF1A was reexpressed after treatment with 5-aza-CdR in a colon cancer cell line
	20% (25/122)	[189]	Sporadic colorectal cancer; data suggest that folate and alcohol intake may be associated with changes in promoter hypermethylation.
	16% (n=149)	[110]	Colorectal carcinoma
	2% (n=95)	[110]	Colorectal adenoma; methylation of RASSF1A is a late event: RASSF1A is rarely methylated in adenoma but significantly methylated in colorectal carcinoma (p<0.001). CpG island methylation plays a more important role in proximal colon tumorigenesis rather than in distal colon tumorigenesis(10.7% (n=56) right colon vs. 19,4% (n=93) in the left colon).
	81% (39/48)	[152]	Early flat-type colorectal tumor; RASSF1A methylation was also detected in [illegible] of normal colonic mucosal tissue; K-ras mutation and RASSF1A methylation occurred mutually exclusive
	3% (2/65)	[208]	
Esophageal cancer	52% (25/48)	[103]	SSC; significant correlation between RASSF1A methylation and advanced tumor stage was detected (p=0.009, stage I/II vs. stage III/IV).
Gastric cancer	43% (39/90)	[19]	Inactivation of RASSF1A was correlated with advanced tumor stage
	67% (14/21)	[88]	Epstein-Barr virus-positive carcinoma
	4% (2/56)	[88]	Epstein-Barr virus-negative carcinoma
	26% (8/31)	[171]	No significant correlation between methylation of RASSF1A and clinicopathological characteristics of the tumors was found. 11% of the gastric intestinal metaplasia samples showed hypermethylation.
Gastric cancer	8% (6/80)	[90]	RASSF1A methylation was found only in gastric cancer. No RASSF1A methylation was observed in gastric adenoma (n=79), intestinal metaplasia (n=57) and chronic gastritis (n=74).
	40%	[74]	Gastrointestinal stromal tumor
	0.4 % (n=268)	[89]	Gastric mucosa samples
Head and neck cancer	8% (6/80)	[65]	
	17% (4/24)	[69]	RASSF1A methylation was higher in poorly differentiated SCC (P<0.005)
	15% (7/46)	[40]	A significant inverse correlation between RASSF1A promoter methylation and HPV infection was found (p=0.038).
	0% (0/32)	[125]	Primary tumors SSC; 26% (5/19) methylation in cancer cell lines

Table 1. RASSF1A methylation in primary human tumors (Contnued)

Tumor entity	Methylation	Ref.	Remarks
Hepatocellular carcinoma	100% (29/29)	[216]	RASSF1A was less frequently methylated in the adjacent non-cancerous liver tissue (24/29, 83%).
	93% (14/15)	[155]	RASSF1A inactivation by methylation is a frequent event in HCC, but was not detected in adenoma.
	95% (41/43)	[217]	The level of methylation in non-tumor tissue was significantly lower than in the corresponding tumor tissue.
	67% (40/60)	[109]	Hepatocellular carcinoma (HCC)
	9% (2/22)	[109]	Dysplastic nodule (DN); no methylation in 30 liver cirrhosis (LC) and 34 chronic hepatitis (CH): HCC vs.DN $p>0.001$; HCC vs. LC $p<0.001$; no correlations with age, sex, stage, survival time.
Kidney cancer	56% (18/32)	[214]	Renal cell carcinoma
	91% (39/43)	[42]	Ectopic re-expression of RASSF1A suppressed growth in vitro
	23% (32/138)	[132]	clear cell renal cell carcinoma; RASSF1A was re-expressed after treatment with 5-aza-CdR in cancer cell lines.
	44% (12/27)	[132]	Papillary renal cell carcinoma
	52% (26/50)	[13]	Kidney tumor; RASSF1 methylation was detected in urine DNA
	100% (6/6)	[13]	Papillary kidney tumor; association of RASSF1A hypermethylation and papillary tumors was statistically significant ($p=0.022$)
	44% (19/43)	[13]	Non-papillary renal cell carcinoma
	45%	[43]	Kidney tumors; RASSF1A methylation was detected at a significant higher frequency in papillary tumors ($p=0.011$) and in high grade tumors ($p=0.003$).
	46% (23/50)	[43]	Clear cell renal cell carcinoma
	70% (14/20)	[43]	Papillary kidney tumor
Kidney cancer	17% (1/6)	[43]	Chromophobe kidney tumor
	14% (1/7)	[43]	Oncocytoma
	60% (3/5)	[43]	Kidney tumor of the collecting duct
	33% (2/6)	[43]	Transitional cell carcinoma of the renal pelvis
	40% (20/50)	[173]	Clear cell renal cell carcinoma (RCC); RASSF1A methylation was found in 84% of RCC cell lines
	100% (9/9)	[62]	Papillary kidney tumor; high level of RASSF1A methylation detected
	90% (19/21)	[62]	Clear cell renal cancer
	25% (2/8)	[62]	Oncocytoma
Lung cancer	38% (22/58)	[30]	Non-small cell lung cancer (NSCLC); exogenous expression of RASSF1A inhibited growth of lung cancer cells in vitro and in vivo
	28% (7/25)	[30]	Adenocarcinoma
	58% (8/14)	[30]	Large cell carcinoma
	37% (7/19)	[30]	Squamous cell carcinoma
	79% (22/28)	[31]	Small cell lung cancer (SCLC)
	72% (21/29)	[1]	SCLC
	34% (14/41)	[1]	NSCLC
	30% (32/107)	[18]	NSCLC; methylation of RASSF1A was associated with impaired patient survival ($P=0.046$)
	32% (35/110)	[174]	RASSF1A methylation correlated with adverse survival of lung adenocarcinoma patients

Table 1. RASSF1A methylation in primary human tumors (Contnued)

Tumor entity	Methylation	Ref.	Remarks
Lung cancer	71%	[181]	Atypical carcinoids.
	45%	[181]	Typical carcinoids ; methylation frequency of RASSF1A was significantly higher in neuroendocrine tumors than in the NSCLC tumors(p<0.0001); methylation of RASSF1A was higher in SCLC tumors than in bronchial carcinoids (p=0.002)
	36% (107/299)	[183]	Adenocarcinoma; no significant differences in methylation status of RASSF1A between smokers and non-smokers
	37% (72/194)	[183]	Squamous cell carcinoma
	21% (5/24)	[70]	No correlation between tumor stage, location and RASSF1A methylation status in sputum samples of NSCLC patients. 50% (n=8) SCLC showed methylation in sputum.
	42% (42/100)	[47]	In the cases of stage I and II diseases RASSF1A methylation was associated with earlier recurrence (p=0.0247).
	32% (66/204)	[96]	Hypermethylation of the RASSF1A promoter was found to be significantly associated with the age of starting smoking (p=0.001); RASSF1A promoter was found to be associated with a poor prognosis in NSCLC patients at stages 1 and 2 (p=0.02 and 0.01, respectively).
	33% (80/242)	[97]	NSCLC; RASSF1A methylation was not associated with K-ras mutations (p=0.37); RASSF1A methylation more frequently in adenocarcinoma (30%) than in squamous cell carcinoma (26%); the hazard of failure for those with RASSF1A methylation was higher compared with that of those with neither K-ras mutation nor RASSF1A methylated (p=0.01).
	43% (32/75)	[210]	Methylation of RASSF1A was cancer-specific (p<0.05).
	45%	[115]	NSCLC; the results indicate a trend of inverse relationship between K-ras activation and RASSF1A promoter methylation
	55%	[115]	Adenocarcinoma
	25%	[115]	Large cell carcinoma
	25%	[115]	Squamous cell carcinoma
	34% (17/50)	[144]	NSCLC; 34% in serum of patients, correlation between methylation in tumor and serum was observed (p=0.0001)
	47% (7/15)	[144]	Adenocarcinoma
	40% (4/10)	[144]	Large cell carcinoma
	24% (6/25)	[144]	Squamous cell carcinoma
	30% (32/107)	[218]	NSCLC; additional analysis of bronchial brushes (6%), bronchoalveolar lavage (5%) and oropharyngeal brushes (4%); methylation events more often in smokers
	41% (51/124)	[128]	NSCLC
	45% (14/31)	[179]	NSCLC; methylation detected in 29% (4/14) bronchoalveolar lavage of tumor patients
	52% (12/21)	[179]	Adenocarcinoma
	28% (17/61)	[76]	Adenocarcinoma
	39% (46/119)	[196]	RASSF1A methylation is a significant predictor of a poor 5-year overall survival rate (p<0.0001)

Table 1. RASSF1A methylation in primary human tumors (Contnued)

Tumor entity	Methylation	Ref.	Remarks
Lung cancer	52% (50/96)	[124]	Adenocarcinoma; RASSF1A methylation was significantly related to starting smoking under age 18
	31% (22/61)	[124]	Squamours cell carcinoma
	39% (28/72)	[175]	Adenocarcinoma; RASSF1A methylation was more frequently observed in AC than in SCC
	13% (6/45)	[175]	Squamours cell carcinoma
	38% (80/209)	[39]	Adenocarcinoma
Lymphoma	65% (34/52)	[134]	Hodgkin's lymphoma; 83% (5/6) in non-Hodgkin lymphoma cell lines; hypermethylation in serum samples: 9% (2/22)
Melanoma	55% (24/44)	[168]	Malignant cutaneous melanoma
	41% (18/44)	[168]	Region upstream from exon 1α of RASSF1A
	50% (22/44)	[168]	Region within exon 1α of RASSF1A
	15% (3/20)	[72]	Primary tumors; hypermethylation of RASSF1A increases during tumor progression
	57% (49/86)	[72]	Metastatic tumors; RASSF1A methylation in 19% (n=6) of plasma from preoperative blood specimen
	53%	[147]	Transcriptional downregulation of RASSF1A does not function as an alternative mechanism to oncogenic BRAF or N-ras mutation in melanomas
	36% (9/25)	[58]	
Mesothelioma	32% (21/66)	[180]	Malignant mesothelioma; inactivation of RASSF1A was correlated with the presence of SV40 in mesothelioma (P=0.022).
Myeloma	28% (9/32)	[139]	Multiple myeloma; no mutation of RASSF1A and BRAF
	15% (17/113)	[157]	Multiple myeloma
	14% (4/29)	[157]	Monoclonal gammopathie of undetermined significance
	0% (0/56)	[59]	Multiple myeloma; 80% in multiple myeloma cell lines
Nasopharyngeal carcinoma	67% (14/21)	[122]	Nasopharyngeal (NP) cancer no significant correlation between methylation of RASSF1A and clinical parameters
	50% (8/16)	[178]	RASSF1A methylation was detected in 39% of EBV associated NP brushing samples
	83% (24/29)	[106]	
	67% (20/30)	[22]	Tumor tissue; methylation of RASSF1A in nasopharyngeal swaps (33%), mouth and throat rinsing fluid (37%) and peripheral blood (3%)
	46% (14/30)	[202]	Undifferentiated NP, methylation of RASSF1A in peripheral blood was detected in all samples with methylated tumor
	65%	[203]	Methylation of RASSF1A was detected in 5% (2/41) of serum of patients with nasopharyngeal carcinoma; the plasma DNA concentration was higher in NPC patients than in normal individuals (p=0.175); hpermethylated gene levels in plasma of NPC patients were not correlated with sex, clinical tumor staging, and lymph node status.
Osteosarcoma	40% (4/10)	[117]	RASSF1A not expressed in 83% (5/6) cell lines; treatment of cell lines with 5-aza-2-deoxycytidine reactivated the transcription of RASSF1A, but not that of RASSF1B

Table 1. RASSF1A methylation in primary human tumors (Contnued)

Tumor entity	Methylation	Ref.	Remarks
Ovarian cancer	10% (2/21)	[1]	
	40% (8/20)	[214]	
	41% (20/49)	[146]	RASSF1A methylation frequency was significantly higher in sporadic ovarian cancer compared to nonmalignant tissue (P=0.01).
	36% (9/25)	[38]	
	50% (25/50)	[36]	Tumor specific methylation of RASSF1A was observed in serum, plasma and peritoneal fluid from cancer patients
	36% (9/25)	[38]	Granulosa cell tumors
Pancreatic carcinoma	64% (29/45)	[33]	Pancreatic adenocarcinomas with K-ras mutation have significantly less RASSF1A methylation and vice versa (p=0.001); methylation was detected in 44% (8/18) of pancreatitis cases
	83% (10/12)	[33]	Endocrine tumors
	75% (36/48)	[75]	Endocrine tumors (ET); tumors larger than 5 cm and those associated with lymph node or hepatic metastases exhibited a higher frequency of methylation at RASSF1A compared with ET's without malignant features
Pediatric tumors	40% (70/175)	[64]	42% in Wilms: tumor, 88% in medulloblastoma, 59% in retinoblastoma61% rhabdomyosarcoma, 52% neuroblastoma, 19% hepatoblastoma, 18% acute leukemia
	73% (22/30)	[46]	Wilms' tumor
	[illegible]	[illegible]	Wilms' tumor
	67% (16/24)	[200]	Including neuroblastoma, thyroid cancer, hepatocellular carcinoma pancreatoblastoma, adrenocortical carcinoma, Wilms' tumor, Burkitt's lymphoma and T-Cell lymphoma; RASSF1A methylation was detected in 54%, 40% and 9% of buffy coat samples before, during and after treatment
Phaeochromo-cytoma	22% (5/23)	[8]	
	48% (12/25)	[35]	RASSF1A methylation was more common in hereditary tumors (58%) compared to the sporadic tumors (38%)
Prostate cancer	53% (54/101)	[127]	RASSF1A methylation was correlated with clinicopathological features of poor prognosis
	100% (11/11)	[104]	Reintroduction of RASSF1A suppressed the growth of a prostate cancer cell line in vitro
Prostate cancer	71% (37/52)	[120]	RASSF1A methylation frequency was higher in more aggressive tumors (p=0.032)
	84% (31/37)	[91]	The methylation frequency of RASSF1A was higher in prostate cancer with high serum PSA (prostate specific antigen) or with high GS(Gleason score) than those with low PSA or GS (p<0.05).
	66% (59/90)	[205]	No correlation between RASSF1A hypermethylation and tumor grade or stage or race of investigated patients was observed; in benign prostate hyperplasia (n=7) no RASSF1A methylation was found
	83% (20/24)	[204]	Frequency of methylation did not differ by tumor grade; 30% (3/10) of RASSF1A methylation was detected in high-grade prostatic intraepithelial neoplasia

Table 1. RASSF1A methylation in primary human tumors (Contnued)

Tumor entity	Methylation	Ref.	Remarks
Prostate cancer	49%	[162]	RASSF1A methylation was found in 19% of benign prostatic hyperplasia
	78% (88/113)	[52]	
	99% (116/117)	[82]	Methylation of RASSF1A was also detected in 93% of benign hyperplasia (BPH) and 100% of intraepithelial neoplasia (HGPIN); methylation levels were significantly higher in prostate cancer than HGPIN and BPH and correlate with advanced tumor stage (p=0.0025)
Soft tissue sarcoma	20% (7/84)	[156]	RASSF1A methylation was more frequent in leiomyosarcoma (39%) compared to malignant fibrous histiocytomas (6%) and liposarcomas (39%); tumor related death of cancer patients with methylated RASSF1A was significantly increased (p=0.037)
Testicular germ cell tumor	22% (20/92)	[100]	80% (8/10) of methylated tumors showed lack or down-regulation of RASSF1A expression.
	40% (4/10)	[71]	Seminomas
	83% (15/18)	[71]	Nonseminomas; RASSF1A methylation was significantly less in seminomas compared to nonseminomas (p=0.0346). RASSF1A methylation occurs early in tumorigenesis
	0% (0/25)	[93]	Testicular germ cell tumors; 100% (3/3) RASSF1A methylation in testicular malignant lymphomas
	36%	[101]	Nonseminoma; 52% in cisplatin resistant tumors vs. 28% in cisplatin sensitive tumors; RASSF1A may serve as a marker for cisplatin resistance; evidence for hypermethylation by cisplatin treatment
Thyroid cancer	71% (27/38)	[154]	RASSF1A methylation was more frequent in more aggressive thyroid carcinomas
	44% (4/9)	[207]	Follicular adenomas
	75% (9/12)	[207]	Follicular thyroid cancer
	20% (6/30)	[207]	Papillary thyroid cancer; in tumor cell lines and in PTCs an inverse correlation of RASSF1A methylation and BRAF mutation was found

When quantitative methylation analysis was applied to determine the methylation level of RASSF1A in breast, prostate and thyroid tissue no significant methylation was detected in non-neoplastic tissue [111, 207, 213]. Recently, hypermethylation of several cancer-related genes (e.g. p16, MGMT, DAP-kinase, RASSF1A, COX2 and RARß) was detected in histologically negative bronchial margins of resected NSCLC [63]. This hypermethylation may represent a field defect of preneoplastic changes that occurs early in carcinogenesis or may be related to aging [77, 195].

Hypermethylation of the RASSF1A promoter and other tumor-related CpG islands were correlated with the exposure to smoke in lung cancer and an earlier age of starting smoking [96, 124, 183, 184]. The effects of dietary folate and alcohol intake on promoter methylation were investigated in patients with sporadic colorectal cancer [189]. Folate supplies a methyl group to convert homocysteine to methionine, which is then converted to Sadenosylmethionine, the methyl donor for a wide variety of biological substrates. Van Engeland et al. (2003) observed that the frequency of RASSF1A methylation is higher (25%) in cancer patients with low methyl donor dietary supplementation (low folate/high alcohol) compared to patients with high folate and low alcohol intake (15%). However, this difference was not significant for RASSF1A and several other tumor suppressor genes [189]. In summary, RASSF1A methylation is one of the most frequent alterations detected in human tumors and may play crucial roles in the initiation, promotion and progression of cancer originating from different tissue.

RASSF1A Methylation as Prognostic Marker

Hypermethylation of RASSF1A occurs frequently in different tumor entities and therefore, aberrant RASSF1A promoter methylation is being widely studied as a biomarker for the prognosis of cancer patients. Various publications have demonstrated that the frequency of RASSF1A hypermethylation in different cancer entities is correlated with clinicopathological aspects, including higher grade of tumors or a reduced time of survival (Table 1). An association of hypermethylation of the promoter of RASSF1A with an advanced tumor stage was found in bladder cancer [108] and in gastric cancer [19]. An increasing RASSF1A methylation frequency from grade II to grade IV glioblastoma was reported [68]. In neuroblastoma, RASSF1A methylation was significantly associated with age >1 ($p<0.01$), high-risk disease ($p<0.016$) and poor survival ($p<0.001$)[212]. Kuroki et al. (2003) have demonstrated a significant correlation between RASSF1A hypermethylation and advanced tumor stage in esophageal squamous cell carcinoma (Table 1). In kidney cancer, RASSF1A methylation was correlated with high grade and papillary tumors [43]. In prostate cancer, RASSF1A methylation was correlated with high serum PSA and high Gleason score [91, 120]. A recent study showed that the methylation level of RASSF1A is correlated with pathologic tumor stage in prostate cancer ($p=0.0025$) [82]. A poorer prognosis of cancer patients with aberrant RASSF1A was reported in bladder cancer [126] and in prostate cancer [127]. In NSCLC, several studies have significantly associated RASSF1A methylation with poor prognosis (stage 1 and stage 2) and advanced tumor stage [18, 47, 96, 174, 196]. RASSF1A methylation was dominantly detected in lung tumors with vascular invasion or pleural involvement and was observed more frequently in poorly differentiated tumors than in well or moderately differentiated tumors [174]. RASSF1A methylation was more frequently observed in adenocarcinoma than in squamous cell carcinoma [124, 175]. It was reported that RASSF1A methylation was associated with an earlier age at starting smoking [96, 124]. A direct correlation between RASSF1A methylation and an earlier recurrence in NSCLC was reported by Endoh et al. (2003). It was observed that the methylation frequency was higher in metastases or in late states of certain cancer (Table 1). In lymph node metastasis of breast tumors, RASSF1A methylation was detected more frequent than in the primary breast carcinoma and hypermethylation was found in metastases in bone, brain and lung [131]. Schagdarsurengin et al. (2002) reported that RASSF1A methylation was more often in undifferentiated and medullary thyroid carcinomas. Müller et al. (2003) detected impaired outcome for breast cancer patients with methylation of the RASSF1A promoter in serum.

Methylation of RASSF1A occurred more frequently in unaffected women at high-risk for breast cancer (Gail model) than in low/intermediate risk women (70% and 29%, respectively; p=0.04) [114]. The risk of a tumor-related death for soft tissue sarcoma patients with methylated RASSF1A was significantly increased [156]. Lee et al. (2004) reported that methylation of the RASSF1A promoter is a late event in colorectal neoplasia. In melanoma, hypermethylation of RASSF1A promoter increased during tumor progression and was more frequent in metastatic melanomas [72]. In contrast to these results, in testicular germ cell tumors [71] and in hepatocellular carcinoma [216] hypermethylation of the promoter of RASSF1A is an early event in tumorigenesis. In testicular germ cell tumors, an association between the resistance towards the chemo-therapeutic agent cisplatin and RASSF1A hypermethylation was observed [101]. Taken together, RASSF1A methylation was often correlated with an advanced tumor stage and poor survival in different tumor entities.

RASSF1A Methylation as a Biomarker for Cancer Patients

The detection of tumors at early stages requires new approaches for characterization and identification of cancer-specific biomarkers and the establishment of reliable non–invasive methods for the detection of these biomarkers in body fluids. Methylation-specific PCR (MSP) has been used in several pilot studies to amplify cancer cell DNA obtained from bodily fluids and these DNA methylation analyses may serve as a powerful new tool for cancer diagnosis [186]. For example DNA isolated from serum of cancer patients was used to screen for tumors in the liver [199], in the lung [48] and for head and neck cancer [153]. Additionally, aberrant methylation of tumor-related genes was detected in DNA obtained from sputum or bronchial lavages of lung cancer patients [3, 14] and urine of prostate cancer [20] and bladder cancer patients [21, 45].

Several studies have analyzed the hypermethylation of the RASSF1A promoter in distinct bodily fluids. In nasopharyngeal carcinoma, Wong et al. (2003) found a RASSF1A methylation frequency of only 5% using MSP out of sera, whereas 65% of the primary tumors showed hypermethylation. A similar frequency of hypermethylation (3%) of RASSF1A in peripheral blood of nasopharyngeal carcinoma patients was detected by Chang et al. (2003), and additionally hypermethylation of RASSF1A was detected in nasopharyngeal swaps (33%) and mouth and throat rinsing fluids (37%) of cancer patients [22]. In 4 out of 14 (29%) lung cancer patients who showed hypermethylation of RASSF1A, the bronchoalveolar lavages were positive for hypermethylated RASSF1A [179]. Zöchbauer-Müller et al. (2003) found hypermethylation of RASSF1A in bronchial brushes (6%), bronchoalveolar lavages (5%) and oropharyngeal brushes (2%). In serum of patients with NSCLC, Ramirez et al. (2003) detected a high frequency (34%) of RASSF1A methylated DNA and a high correlation between methylation in tumor tissue and serum (p=0.0001). Another study found in 12% of serum DNA from lung cancer patients RASSF1A methylation and the smoking status was positively associated with its methylation [54]. Sputum of lung cancer patients is another body fluid, which was investigated for hypermethylated DNA of tumor suppressor genes [15, 70]. Honorio et al. (2003) found that in 50% of SCLC and in

21% of NSCLC sputum samples a hypermethylation of RASSF1A was detectable. Belinsky et al. (2002) detected only a low frequency of RASSF1A methylation in the sputum of controls. In 50% sera of glioblastoma patients, RASSF1A methylation was observed and this correlated with the RASSF1A inactivation in the tumor tissue [145]. RASSF1A methylation was also investigated in the urine of patients with bladder and kidney cancer [13, 21, 45, 53]. In 19 of 23 (82%) of patients with a RASSF1A methylated bladder cancer, Dulaimi et al. (2004) detected RASS1A hypermethylation in the urine samples. MSP used for detection of a panel of methylated promoters of cancer related genes (APC, RASSF1A and p14) in urine of bladder cancer patients showed 100% specificity and methylation of RASSF1A was not detected in the urine samples of controls [45]. Chan et al. (2003) detected methylation of RASS1A in 7 out of 14 (50 %) urine specimen and all positive probes showed also epigenetic inactivation of RASSF1A in the corresponding primary bladder tumors. In urine samples of patients with kidney tumor, Battagli et al. (2003) found in 25 out of 50 (50%) hypermethylated RASSF1A. Only for a single case, no RASSF1A methylation was detected in the urine despite of a methylated tumor [13]. However, a recent study indicates the RASSF1A is frequently methylated in urine sediments of cancer free individuals [53]. In patients with Hodgkins' lymphoma, hypermethylation of the RASSF1A promoter occurred in two out of 22 (9%) sera, whereas the frequency of hypermethylation in primary tumors was 65% [134]. In 22 samples of breast cancer, a panel of six genes (GSTP1, RARß2, p16, p14, RASSF1A and DAP-kinase) was examined by Krassenstein et al. (2004), and in the corresponding nipple aspirates. At least one gene showed hypermethylation in its promoter region in the tumor samples and methylation of the same gene was detected in 18 out of 22 (82%) nipple aspirates [102]. In sera of breast cancer patients, hypermethylation of cancer-related genes was investigated and methylation of RASSF1A was only detected in serum DNA of cancer patients with a methylated RASSF1A CpG island [44, 133]. Müller et al. (2003) reported that methylation of the RASSF1A promoter was detected in 6 out of 26 (23%) and associated with a worse prognosis and a poor outcome [133]. Recently, hypermethylation of RASSF1A was also detected in tampons of patients with endometrial cancer [51]. RASSF1A methylation was detected in bodily fluids (serum, plasma and peritoneal fluid) of ovarian cancer patients with 100% specificity [36]. Methylation was undetectable in bodily fluids of patients with RASSF1A-unmethylated tumors and in controls. Taken together, RASSF1A hypermethylation is frequently detected in bodily fluids of cancer patients. Different frequencies in DNA methylation could be attributed to various DNA concentrations of disseminating cancer cells in serum, sputum and other bodily fluids compared to the primary tumors and limitations in the sensitivity of the detection system. Methylation analysis of tumor-related genes in easily obtainable bodily fluids is a promising new experimental approach to screen putative cancer patients.

Function of RASSF1A as Tumor Suppressor

RASSF1A is involved in several growth regulating and apoptotic pathways and regulates cell proliferation, cellular integrity and cell death (Figure 3). Ectopic expression of RASSF1A in cancer cell lines, which lack endogenous RASSF1A transcription resulted in reduced colony formation and/or anchorage-independent growth in soft agar in lung, kidney, prostate, glioma and nasopharyngeal cancer cell lines [18, 26, 30, 42, 68, 104, 116]. In nude mice, human cancer cells lacking RASSF1A transcription formed larger tumors compared to the same cells expressing exogenous RASSF1A [18, 26, 30, 116]. Mutant RASSF1A had only a reduced growth suppression activity *in vivo* and *in vitro* [42, 116]. Ectopic expression of the *RASSF1C* isoform showed only a modest reduction of cell viability *in vitro* [83]. However, a recent report indicates that in a renal cancer cell line overexpression of RASSF1C inhibits growth and induces cell cycle arrest [116]. To gain insight into RASSF1A function, expression profiles of cancer cell lines, which re-expressed RASSF1A were generated [2]. Agathanggelou et al. (2004) have characterized several genes (e.g ETS2, Cyclin D3, CDH2, DAPK1, TXN and CTSL) that may represent gene expression targets for RASSF1A. Shivakumar et al. (2002) have reported that RASSF1A can induce cell-cycle arrest by engaging the Rb family cell-cycle checkpoint. E7 papilloma virus protein-expressing cells are resistant to the RASSF1A-induced cell-cycle arrest [161]. RASSF1A also inhibits accumulation of native cyclin D1 (Figure 3) and the RASSF1A-induced growth arrest can be relieved by ectopic expression of cyclins, but not by oncogenic Ras expression [161].

Activated Ras is usually associated with enhanced proliferation, transformation and cell survival (Figure 3). Ras also induces proliferation inhibitory effects [12, 159] and apoptosis [24, 41, 130, 160]. Ras effectors, like RASSF1A, may be specialized to inhibit cell growth and to induce cell death and these inhibitory signaling pathways may need to be inactivated during carcinogenesis. Vos et al. (2000) have shown that RASSF1C binds Ras in a GTP-dependent manner and expression of RASSF1C induced apoptosis. This pro-apoptotic effect of RASSF1 is enhanced by activated Ras and inhibited by dominant negative Ras [191]. Recent data indicate that in colorectal and pancreatic cancer, the inactivation of RASSF1A and activation of Ras are mutual exclusive [33, 152, 188], but in lung cancer this correlation was not significant [97, 144, 174]. In thyroid cancer, RASSF1A methylation occurred significantly when BRAF was not mutated [207].

Murine models of human cancer may expedite our understanding of carcinogenesis and Rassf1a knockout mice may help to dissect the tumorigenic process involved in the function of Rassf1a. Smith et al. (2002) have created a mouse with a 370 kb deletion of the region homologue to the 3p21.3, which includes Rassf1a. The homozygous deletion of this region is embryonic lethal in mouse [164]. Recently, we have generated Rassf1a-specific knockout mice and consistent with the tumor-suppressive role of RASSF1A, we have observed that these animals were prone to spontaneous and induced carcinogenesis [177]. Interestingly, heterozygous and homozygous Rassf1a knockout mice were significantly more susceptible to spontaneous tumorigenesis ($p<0.05$ and $p<0.001$, respectively) [177]. When heterozygous and homozygous knockout mice were treated with two chemical carcinogens (benzopyrene and urethane), the Rassf1a deficient mice showed increased tumor multiplicity and tumor size compared to the controls [177]. Functional data indicate that Rassf1a knockout embryonic fibroblasts are more sensitive to induced microtubule instability relative to wildtype cells [121]. These functional data and the Rassf1a knockout mice support the tumor suppressor role of RASSF1A observed in cancer.

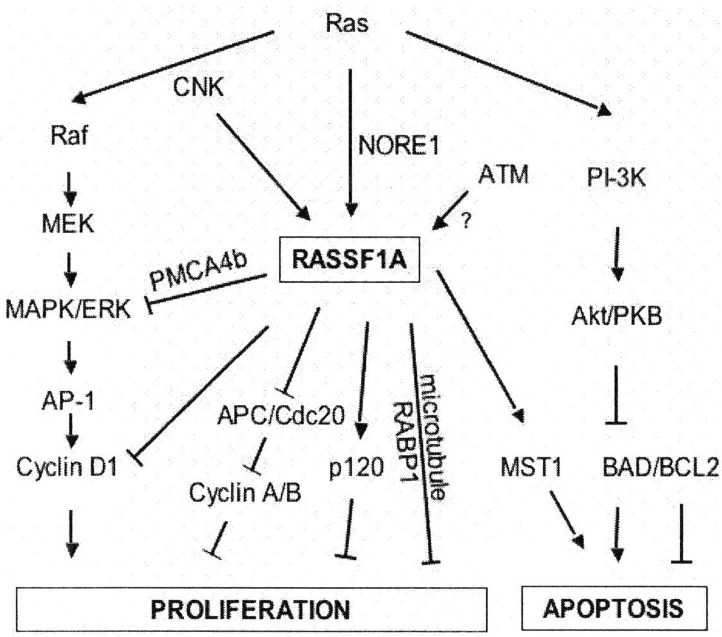

Figure 3. Signaling pathway of RASSF1A-mediated functions. Ras activates cellular proliferation through the MAP-kinase pathway (Raf, MEK, and MAPK/ERK). RASSF1A regulates cellular integrity and proliferation through its interaction with the microtubules, RASSF1A binding protein 1 (RABP1) and CDC20 by inhibiting the anaphase promoting complex and the degradation of cyclin A and B. RASSF1A inhibits the epidermal growth factor dependent activation of Erk through the plasma membrane calmodulin-dependent calcium ATPase 4b (PMCA4b). RASSF1A inhibits the accumulation of cyclin D1 and interacts with the transcription factor p120E4F. Ras inhibits apoptosis by the pathway of the phosphatidylinositol 3 kinase (PI3-K) and stimulates the activity of the protein kinase B Akt (Akt/PKB). Subsequently, Akt inhibits apoptosis induced by members of the Bcl-family (BAD). The RASSF1A tumor suppressor induces apoptosis through its interaction with Ras, the novel Ras effector (NORE1) the connector enhancer of KSR (CNK) and the pro-apoptotic MST1 kinase. The exact role of the ataxia telangiectasia mutated (ATM) kinase is unclear.

Another homologue of RASSF1, which encodes a Ras association domain, was characterized in mouse and human and was termed novel Ras effector (NORE1) [176, 190]. Recent data show that the RASSF1A-related Ras effector NORE1 may serve as a Ras-regulated tumor suppressor in lung cancer and melanoma [6, 192] and epigenetic inactivation of NORE1 was detected in several cancers, including lung and kidney cancer [25, 67, 176]. No correlation between RASSF1A methylation and NORE1 inactivation was reported [67], however, in lung cancer hypermethylation of NORE1 occurs preferentially in the context of a wild-type K-ras [76]. Our results indicate that binding of RASSF1A to Ras may require heterodimerization with NORE1, and that RASSF1A binds to Ras only weakly by itself [140]. RASSF1A and NORE1 may function in the same Ras-regulated pathway. Khokhlatchev et al. (2002) showed that RASSF1A and NORE1 interact with the pro-apoptotic kinase MST1, which mediates the apoptotic effect of activated Ras. MST1 is a member of the group II germinal center (protein serine/threonine) kinases and the NORE1/RASSF1-MST1 complex represents a novel Ras-regulated proapoptotic pathway

[94]. Praskova et al. (2004) have reported that the MST1 kinase is regulated by robust auto-activation, which is mediated by auto-phosphorylation. Co-expression of RASSF1 and NORE1 suppressed the phosphorylation and therefore the auto-activation of MST1 [142]. Moreover, MST1 activity is stimulated by membrane recruitment and when bound to Ras. Recently, Rabizadeh et al. (2004) have reported that the scaffold protein CNK1 interacts with the tumor suppressor RASSF1A and augments RASSF1A induced cell death. The connector enhancer of KSR (CNK) is a c-Raf1 binding protein, which mediates Ras-induced Raf activation. CNK1 is an interaction partner of RASSF1 and represses growth of dividing cancer cells and initiates apoptosis through the MST1 (or MST2) pathway (Figure 3). However, RASSF1C does not influence CNK1 induced apoptosis [143]. In addition to a pro-proliferating role of CNK1 by activated Ras, CNK1 also participates in a pro-apoptotic pathway through its binding of the RASSF1A-MST complex [143].

In a yeast two-hybrid screen, RASSF1A was identified as a novel interaction partner of PMCA4b, a plasma membrane calmodulin-dependent calcium ATPase [7]. The functionality of the interaction was demonstrated by inhibition of the epidermal growth factor-dependent activation of the Erk pathway when PMCA4b and RASSF1 were coexpressed (Figure 3). In another screen, p120E4F was identified as an interaction partner of RASSF1A [50]. p120E4F is an E1A-regulated transcription factor which interacts with the retinoblastoma protein, p14ARF and p53 and is involved in control of cell cycle arrest near the G1 transition. The G1 cell cycle arrest and S phase inhibition was enhanced by p120E4F in the presence of RASSF1A [50].

Several different groups have reported that RASSF1A is a microtubule-binding protein, which regulates mitotic progression [29, 119, 121, 149, 150, 165, 193]. We have shown that RASSF1A co-localizes with microtubules in interphase and decorates spindles and centrosomes during mitosis [121]. Upon binding to microtubules, RASSF1A has a strong cytoprotective activity against microtubule depolymerization induced by nocodazole *in vivo*. RASSF1A-/- cells were more sensitive against nocodazole induced G2/M arrest than wild type cells [121]. The domain required for both microtubule association and stabilization was mapped to a fragment that contains the Ras association domain. Overexpression of RASSF1A induced mitotic arrest at metaphase with aberrant mitotic cells. These results were confirmed by several other groups [29, 119, 149, 150, 165, 193]. RASSF1 was identified as an interaction partner of the C19ORF5 protein with high similarity to microtubule-associated proteins (MAP1A and MAP1B) [29, 119]. Recently C19ORF was redesignated RASSF1A-binding protein 1 (RABP1) and contains two microtubule-associated protein domains [166]. RABP1 plays an important role in the recruitment of RASSF1A to the spindle poles and in the subsequent RASSF1A-mediated regulation of the APC in mitosis [166]. Dallol et al. (2004) have found that RASSF1A substitutions at codon 65 and 257 perturb the association with microtubules and these mutants are less potent inhibitors of DNA synthesis compared to the wildtype protein. Vos et al. (2004) have shown that a deletion mutant of RASSF1A, which lacks the microtubule association domain of RASSF1 is severely defective for the ability to promote cell cycle arrest and partially inhibits RASSF1A induced cell death. Interestingly, it was also shown that wild type RASSF1A and RASSF1C inhibit genomic instability induced by activated Ras [193]. RASSF1A can regulate microtubule stability and induces G2/M arrest [149, 150]. Song et al. (2004) have reported that RASSF1A regulates mitosis by inhibiting the anaphase promoting complex (APC) through Cdc20 and induces G2-M arrest at pro-metaphase. RASSF1C had no effect on the APC-Cdc20 complex and cell cycle regulation [165]. The N-terminal region of RASSF1A interacts directly with Cdc20; however the C-terminus, which encodes the microtubule association domain is also involved in the inhibition of APC [165]. After interaction with RASSF1A, Cdc20 is inhibited to activate APC and therefore APC is unable to degrade the mitotic cyclins A and B (Figure 3). Inactivation of RASSF1A resulted in the acceleration of mitotic progression and in premature destruction of cyclin A and B [165]. The function of RASSF1A is independent of the protein Emi1 (early mitotic inhibitor 1) and therefore Song et al. (2004) proposed that RASSF1A acts in early prometaphase, before activation of the spindle checkpoint and after Emi1 destruction to prevent the degradation of mitotic cyclins and to delay mitotic progression. Thus, RASSF1A may function as a crucial link between tumor suppression and mitotic cell division through a new mechanism [79, 129]. In summary, RASSF1A induces apoptosis and inhibits cellular proliferation through several pathways (Figure 3). RASSF1A can regulate microtubule stability and control mitotic progression, presumably by modulating centrosomes and spindle function and by regulating the APC complex and accumulation of cyclins A, B and D1.

Conclusion

Epigenetic silencing of the RASSF1A tumor suppressor was found in a variety of primary human tumors including, bladder, brain, breast, cervix, colorectal, gastric, liver, kidney, lung, nasopharyngeal, ovarian, pancreatic, prostate and thyroid cancer. RASSF1A silencing was significantly associated with an advanced tumor stage and poorly differentiated tumors. Moreover, hypermethylation of RASSF1A correlated with poor prognosis and impaired survival of the cancer patient. Methylation analysis of the RASSF1A gene could serve as the basis of a diagnostic test for cancer. The observation that RASSF1A inactivation and K-ras activation are mutually exclusive events in the development of certain carcinomas could further pinpoint the function of RASSF1A as a negative effector of Ras. RASSF1A functions as *bona fide* tumor suppressor gene in cancer through several distinct pathways, including apoptosis, genomic and microtubule stability and cell cycle regulation. The exact mechanism of its biological function is complex and may require additional Ras effectors like NORE1. Understanding the molecular function of RASSF1A may lead to the development of new anticancer drugs and the detection of the aberrant methylation of RASSF1A in patients bodily fluids may serve as promising biomarker for cancer diagnosis. Since *RASSF1A* blocks cell cycle progression the silencing of *RASSF1A* may be a critical step in tumorigenesis. Inactivation of *RASSF1A* may serve as a model for epigenetic CpG island silencing of tumor suppressor genes in cancer pathogenesis. It will be interesting to analyze the influence of inhibitors of DNA methylation and histone deacetylase on the reactivation of the *RASSF1A* gene in silenced cells.

Acknowledgements

Data presented in this review were supported by BMBF grant FKZ 01ZZ0104, Land Sachsen-Anhalt, DFG grant DA552-1 to Reinhard Dammann.

References

[1] Agathanggelou, A; Honorio, S; Macartney, DP; Martinez, A; Dallol, A; Rader, J; Fullwood, P; Chauhan, A; Walker, R; Shaw, JA; Hosoe, S; Lerman, MI; Minna, JD; Maher, ER; Latif, F. Methylation associated inactivation of RASSF1A from region 3p21.3 in lung, breast and ovarian tumours. *Oncogene*, 2001, 20, 1509-1518.

[2] Agathanggelou, A; Bieche, I; Ahmed-Choudhury, J; Nicke, B; Dammann, R; Baksh, S; Gao, B; Minna, JD; Downward, J; Maher, ER; Latif, F. Identification of novel gene expression targets for the Ras association domain family 1 (RASSF1A) tumor suppressor gene in non-small cell lung cancer and neuroblastoma. *Cancer Res*, 2003, 63, 5344-5351.

[3] Ahrendt, SA; Chow, JT; Xu, LH; Yang, SC; Eisenberger, CF; Esteller, M; Herman, JG; Wu, L; Decker, PA; Jen, J; Sidransky, D. Molecular detection of tumor cells in bronchoalveolar lavage fluid from patients with early stage lung cancer. *J Natl Cancer Inst*, 1999, 91, 332-339.

[4] Ahuja, N; Li, Q; Mohan, AL; Baylin, SB; Issa, JP. Aging and DNA methylation in colorectal mucosa and cancer. *Cancer Res*, 1998, 58, 5489-5494.

[5] Antequera, F; Boyes, J; Bird, A. High levels of de novo methylation and altered chromatin structure at CpG islands in cell lines. *Cell*, 1990, 62, 503-514.

[6] Aoyama, Y; Avruch, J; Zhang, XF. Nore1 inhibits tumor cell growth independent of Ras or the MST1/2 kinases. *Oncogene*, 2004, 23, 3426-3433.

[7] Armesilla, AL; Williams, JC; Buch, MH; Pickard, A; Emerson, M; Cartwright, EJ; Oceandy, D; Vos, MD; Gillies, S; Clark, GJ; Neyses, L. Novel functional interaction between the plasma membrane Ca2+ pump 4b and the proapoptotic tumor suppressor Ras-associated factor 1 (RASSF1). *J Biol Chem*, 2004, 279, 31318-31328.

[8] Astuti, D; Agathanggelou, A; Honorio, S; Dallol, A; Martinsson, T; Kogner, P; Cummins, C; Neumann, HP; Voutilainen, R; Dahia, P; Eng, C; Maher, ER; Latif, F. RASSF1A promoter region CpG island hypermethylation in phaeochromocytomas and neuroblastoma tumours. *Oncogene*, 2001, 20, 7573-7577.

[9] Astuti, D; Da Silva, NF; Dallol, A; Gentle, D; Martinsson, T; Kogner, P; Grundy, R; Kishida, T; Yao, M; Latif, F; Maher, ER. SLIT2 promoter methylation analysis in neuroblastoma, Wilms' tumour and renal cell carcinoma. *Br J Cancer*, 2004, 90, 515-521.

[10] Balana, C; Ramirez, JL; Taron, M; Roussos, Y; Ariza, A; Ballester, R; Sarries, C; Mendez, P; Sanchez, JJ; Rosell, R. O6-methyl-guanine-DNA methyltransferase methylation in serum and tumor DNA predicts response to 1,3-bis(2-chloroethyl)-1-nitrosourea but not to temozolamide plus cisplatin in glioblastoma multiforme. *Clin Cancer Res*, 2003, 9, 1461-1468.

[11] Bannister, AJ; Zegerman, P; Partridge, JF; Miska, EA; Thomas, JO; Allshire, RC; Kouzarides, T. Selective recognition of methylated lysine 9 on histone H3 by the HP1 chromo domain. *Nature*, 2001, 410, 120-124.

[12] Bar-Sagi, D; Feramisco, JR. Microinjection of the ras oncogene protein into PC12 cells induces morphological differentiation. *Cell*, 1985, 42, 841-848.

[13] Battagli, C; Uzzo, RG; Dulaimi, E; Ibanez de Caceres, I; Krassenstein, R; Al-Saleem, T; Greenberg, RE; Cairns, P. Promoter hypermethylation of tumor suppressor genes in urine from kidney cancer patients. *Cancer Res*, 2003, 63, 8695-8699.

[14] Belinsky, SA; Nikula, KJ; Palmisano, WA; Michels, R; Saccomanno, G; Gabrielson, E; Baylin, SB; Herman, JG. Aberrant methylation of p16(INK4a) is an early event in lung cancer and a potential biomarker for early diagnosis. *Proc Natl Acad Sci U S A*, 1998, 95, 11891-11896.

[15] Belinsky, SA; Palmisano, WA; Gilliland, FD; Crooks, LA; Divine, KK; Winters, SA; Grimes, MJ; Harms, HJ; Tellez, CS; Smith, TM; Moots, PP; Lechner, JF; Stidley, CA; Crowell, RE. Aberrant promoter methylation in bronchial epithelium and sputum from current and former smokers. *Cancer Res*, 2002, 62, 2370-2377.

[16] Bird, A. DNA methylation patterns and epigenetic memory. *Genes Dev*, 2002, 16, 6-21.

[17] Bird, AP. CpG-rich islands and the function of DNA methylation. *Nature*, 1986, 321, 209-213.

[18] Burbee, DG; Forgacs, E; Zochbauer-Muller, S; Shivakumar, L; Fong, K; Gao, B; Randle, D; Kondo, M; Virmani, A; Bader, S; Sekido, Y; Latif, F; Milchgrub, S; Toyooka, S; Gazdar, AF; Lerman, MI; Zabarovsky, E; White, M; Minna, JD. Epigenetic inactivation of RASSF1A in lung and breast cancers and malignant phenotype suppression. *J Natl Cancer Inst*, 2001, 93, 691-699.

[19] Byun, DS; Lee, MG; Chae, KS; Ryu, BG; Chi, SG. Frequent epigenetic inactivation of rassf1a by aberrant promoter hypermethylation in human gastric adenocarcinoma. *Cancer Res*, 2001, 61, 7034-7038.

[20] Cairns, P; Esteller, M; Herman, JG; Schoenberg, M; Jeronimo, C; Sanchez-Cespedes, M; Chow, NH; Grasso, M; Wu, L; Westra, WB; Sidransky, D. Molecular detection of prostate cancer in urine by GSTP1 hypermethylation. *Clin Cancer Res*, 2001, 7, 2727-2730.

[21] Chan, MW; Chan, LW; Tang, NL; Lo, KW; Tong, JH; Chan, AW; Cheung, HY; Wong, WS; Chan, PS; Lai, FM; To, KF. Frequent hypermethylation of promoter region of RASSF1A in tumor tissues and voided urine of urinary bladder cancer patients. *Int J Cancer*, 2003, 104, 611-616.

[22] Chang, HW; Chan, A; Kwong, DL; Wei, WI; Sham, JS; Yuen, AP. Evaluation of hypermethylated tumor suppressor genes as tumor markers in mouth and throat rinsing fluid, nasopharyngeal swab and peripheral blood of nasopharygeal carcinoma patient. *Int J Cancer*, 2003, 105, 851-855.

[23] Chen, CM; Chen, HL; Hsiau, TH; Hsiau, AH; Shi, H; Brock, GJ; Wei, SH; Caldwell, CW; Yan, PS; Huang, TH. Methylation target array for rapid analysis of CpG island hypermethylation in multiple tissue genomes. *Am J Pathol*, 2003, 163, 37-45.

[24] Chen, CY; Liou, J; Forman, LW; Faller, DV. Differential regulation of discrete apoptotic pathways by Ras. *J Biol Chem*, 1998, 273, 16700-16709.

[25] Chen, J; Lui, WO; Vos, MD; Clark, GJ; Takahashi, M; Schoumans, J; Khoo, SK; Petillo, D; Lavery, T; Sugimura, J; Astuti, D; Zhang, C; Kagawa, S; Maher, ER; Larsson, C; Alberts, AS; Kanayama, HO; Teh, BT. The t(1;3) breakpoint-spanning genes LSAMP and NORE1 are involved in clear cell renal cell carcinomas. *Cancer Cell*, 2003, 4, 405-413.

[26] Chow, LS; Lo, KW; Kwong, J; To, KF; Tsang, KS; Lam, CW; Dammann, R; Huang, DP. RASSF1A is a target tumor suppressor from 3p21.3 in nasopharyngeal carcinoma. *Int J Cancer*, 2004, 109, 839-847.

[27] Clark, SJ; Melki, J. DNA methylation and gene silencing in cancer: which is the guilty party? *Oncogene*, 2002, 21, 5380-5387.

[28] Cohen, Y; Singer, G; Lavie, O; Dong, SM; Beller, U; Sidransky, D. The RASSF1A tumor suppressor gene is commonly inactivated in adenocarcinoma of the uterine cervix. *Clin Cancer Res*, 2003, 9, 2981-2984.

[29] Dallol, A; Agathanggelou, A; Fenton, SL; Ahmed-Choudhury, J; Hesson, L; Vos, MD; Clark, GJ; Downward, J; Maher, ER; Latif, F. RASSF1A interacts with microtubule-associated proteins and modulates microtubule dynamics. *Cancer Res*, 2004, 64, 4112-4116.

[30] Dammann, R; Li, C; Yoon, JH; Chin, PL; Bates, S; Pfeifer, GP. Epigenetic inactivation of a RAS association domain family protein from the lung tumour suppressor locus 3p21.3. *Nat Genet*, 2000, 25, 315-319.

[31] Dammann, R; Takahashi, T; Pfeifer, GP. The CpG island of the novel tumor suppressor gene RASSF1A is intensely methylated in primary small cell lung carcinomas. *Oncogene*, 2001, 20, 3563-3567.

[32] Dammann, R; Yang, G; Pfeifer, GP. Hypermethylation of the cpG island of Ras association domain family 1A (RASSF1A), a putative tumor suppressor gene from the 3p21.3 locus, occurs in a large percentage of human breast cancers. *Cancer Res*, 2001, 61, 3105-3109.

[33] Dammann, R; Schagdarsurengin, U; Liu, L; Otto, N; Gimm, O; Dralle, H; Boehm, BO; Pfeifer, GP; Hoang-Vu, C. Frequent RASSF1A promoter hypermethylation and K-ras mutations in pancreatic carcinoma. *Oncogene*, 2003, 22, 3806-3812.

[34] Dammann, R; Schagdarsurengin, U; Strunnikova, M; Rastetter, M; Seidel, C; Liu, L; Tommasi, S; Pfeifer, GP. Epigenetic inactivation of the Ras-association domain family 1 (RASSF1A) gene and its function in human carcinogenesis. *Histol Histopathol*, 2003, 18, 665-677.

[35] Dammann, R; Schagdarsurengin, U; Seidel, C; Trumpler, C; Hoang-Vu, C; Gimm, O; Dralle, H; Pfeifer, GP; Brauckhoff, M. Frequent promoter methylation of tumor-related genes in sporadic and men2-associated pheochromocytomas. *Exp Clin Endocrinol Diabetes*, 2005, 113, 1-7.

[36] de Caceres, I; Battagli, C; Esteller, M; Herman, JG; Dulaimi, E; Edelson, MI; Bergman, C; Ehya, H; Eisenberg, BL; Cairns, P. Tumor cell-specific BRCA1 and RASSF1A hypermethylation in serum, plasma, and peritoneal fluid from ovarian cancer patients. *Cancer Res*, 2004, 64, 6476-6481.

[37] Dennis, K; Fan, T; Geiman, T; Yan, Q; Muegge, K. Lsh, a member of the SNF2 family, is required for genome-wide methylation. *Genes Dev*, 2001, 15, 2940-2944.

[38] Dhillon, VS; Aslam, M; Husain, SA. The contribution of genetic and epigenetic changes in granulosa cell tumors of ovarian origin. *Clin Cancer Res*, 2004, 10, 5537-5545.

[39] Divine, KK; Pulling, LC; Marron-Terada, PG; Liechty, KC; Kang, T; Schwartz, AG; Bocklage, TJ; Coons, TA; Gilliland, FD; Belinsky, SA. Multiplicity of abnormal promoter methylation in lung adenocarcinomas from smokers and never smokers. *Int J Cancer*, 2005, 114, 400-405.

[40] Dong, SM; Sun, DI; Benoit, NE; Kuzmin, I; Lerman, MI; Sidransky, D. Epigenetic inactivation of RASSF1A in head and neck cancer. *Clin Cancer Res*, 2003, 9, 3635-3640.

[41] Downward, J. Ras signalling and apoptosis. *Curr Opin Genet Dev*, 1998, 8, 49-54.

[42] Dreijerink, K; Braga, E; Kuzmin, I; Geil, L; Duh, FM; Angeloni, D; Zbar, B; Lerman, MI; Stanbridge, EJ; Minna, JD; Protopopov, A; Li, J; Kashuba, V; Klein, G; Zabarovsky, ER. The candidate tumor suppressor gene, RASSF1A, from human chromosome 3p21.3 is involved in kidney tumorigenesis. *Proc Natl Acad Sci U S A*, 2001, 98, 7504-7509.

[43] Dulaimi, E; De, C, II; Uzzo, RG; Al-Saleem, T; Greenberg, RE; Polascik, TJ; Babb, JS; Grizzle, WE; Cairns, P. Promoter hypermethylation profile of kidney cancer. *Clin Cancer Res*, 2004, 10, 3972-3979.

[44] Dulaimi, E; Hillinck, J; Ibanez de Caceres, I; Al-Saleem, T; Cairns, P. Tumor suppressor gene promoter hypermethylation in serum of breast cancer patients. *Clin Cancer Res*, 2004, 10, 6189-6193.

[45] Dulaimi, E; Uzzo, RG; Greenberg, RE; Al-Saleem, T; Cairns, P. Detection of bladder cancer in urine by a tumor suppressor gene hypermethylation panel. *Clin Cancer Res*, 2004, 10, 1887-1893.

[46] Ehrlich, M; Jiang, G; Fiala, E; Dome, JS; Yu, MC; Long, TI; Youn, B; Sohn, OS; Widschwendter, M; Tomlinson, GE; Chintagumpala, M; Champagne, M; Parham, D; Liang, G; Malik, K; Laird, PW. Hypomethylation and hypermethylation of DNA in Wilms tumors. *Oncogene*, 2002, 21, 6694-6702.

[47] Endoh, H; Yatabe, Y; Shimizu, S; Tajima, K; Kuwano, H; Takahashi, T; Mitsudomi, T. RASSF1A gene inactivation in non-small cell lung cancer and its clinical implication. *Int J Cancer*, 2003, 106, 45-51.

[48] Esteller, M; Sanchez-Cespedes, M; Rosell, R; Sidransky, D; Baylin, SB; Herman, JG. Detection of aberrant promoter hypermethylation of tumor suppressor genes in serum DNA from non-small cell lung cancer patients. *Cancer Res*, 1999, 59, 67-70.

[49] Fackler, MJ; McVeigh, M; Evron, E; Garrett, E; Mehrotra, J; Polyak, K; Sukumar, S; Argani, P. DNA methylation of RASSF1A, HIN-1, RAR-beta, Cyclin D2 and Twist in in situ and invasive lobular breast carcinoma. *Int J Cancer*, 2003, 107, 970-975.

[50] Fenton, SL; Dallol, A; Agathanggelou, A; Hesson, L; Ahmed-Choudhury, J; Baksh, S; Sardet, C; Dammann, R; Minna, JD; Downward, J; Maher, ER; Latif, F. Identification of the E1A-regulated transcription factor p120 E4F as an interacting partner of the RASSF1A candidate tumor suppressor gene. *Cancer Res*, 2004, 64, 102-107.

[51] Fiegl, H; Gattringer, C; Widschwendter, A; Schneitter, A; Ramoni, A; Sarlay, D; Gaugg, I; Goebel, G; Muller, HM; Mueller-Holzner, E; Marth, C; Widschwendter, M.

Methylated DNA collected by tampons--a new tool to detect endometrial cancer. *Cancer Epidemiol Biomarkers Prev*, 2004, 13, 882-888.

[52] Florl, AR; Steinhoff, C; Muller, M; Seifert, HH; Hader, C; Engers, R; Ackermann, R; Schulz, WA. Coordinate hypermethylation at specific genes in prostate carcinoma precedes LINE-1 hypomethylation. *Br J Cancer*, 2004, 91, 985-994.

[53] Friedrich, MG; Weisenberger, DJ; Cheng, JC; Chandrasoma, S; Siegmund, KD; Gonzalgo, ML; Toma, MI; Huland, H; Yoo, C; Tsai, YC; Nichols, PW; Bochner, BH; Jones, PA; Liang, G. Detection of methylated apoptosis-associated genes in urine sediments of bladder cancer patients. *Clin Cancer Res*, 2004, 10, 7457-7465.

[54] Fujiwara, K; Fujimoto, N; Tabata, M; Nishii, K; Matsuo, K; Hotta, K; Kozuki, T; Aoe, M; Kiura, K; Ueoka, H; Tanimoto, M. Identification of epigenetic aberrant promoter methylation in serum DNA is useful for early detection of lung cancer. *Clin Cancer Res*, 2005, 11, 1219-1225.

[55] Fuks, F; Burgers, WA; Godin, N; Kasai, M; Kouzarides, T. Dnmt3a binds deacetylases and is recruited by a sequence-specific repressor to silence transcription. *Embo J*, 2001, 20, 2536-2544.

[56] Fuks, F; Hurd, PJ; Deplus, R; Kouzarides, T. The DNA methyltransferases associate with HP1 and the SUV39H1 histone methyltransferase. *Nucleic Acids Res*, 2003, 31, 2305-2312.

[57] Fuks, F; Hurd, PJ; Wolf, D; Nan, X; Bird, AP; Kouzarides, T. The methyl-CpG-binding protein MeCP2 links DNA methylation to histone methylation. *J Biol Chem*, 2003, 278, 4035-4040.

[58] Furuta, J; Umebayashi, Y; Miyamoto, K; Kikuchi, K; Otsuka, F; Sugimura, T; Ushijima, T. Promoter methylation profiling of 30 genes in human malignant melanoma. *Cancer Sci*, 2004, 95, 962-968.

[59] Galm, O; Wilop, S; Reichelt, J; Jost, E; Gehbauer, G; Herman, JG; Osieka, R. DNA methylation changes in multiple myeloma. *Leukemia*, 2004, 18, 1687-1692.

[60] Gao, Y; Guan, M; Su, B; Liu, W; Xu, M; Lu, Y. Hypermethylation of the RASSF1A gene in gliomas. *Clin Chim Acta*, 2004, 349, 173-179.

[61] Gibbons, RJ; McDowell, TL; Raman, S; O'Rourke, DM; Garrick, D; Ayyub, H; Higgs, DR. Mutations in ATRX, encoding a SWI/SNF-like protein, cause diverse changes in the pattern of DNA methylation. *Nat Genet*, 2000, 24, 368-371.

[62] Gonzalgo, ML; Yegnasubramanian, S; Yan, G; Rogers, CG; Nicol, TL; Nelson, WG; Pavlovich, CP. Molecular profiling and classification of sporadic renal cell carcinoma by quantitative methylation analysis. *Clin Cancer Res*, 2004, 10, 7276-7283.

[63] Guo, M; House, MG; Hooker, C; Han, Y; Heath, E; Gabrielson, E; Yang, SC; Baylin, SB; Herman, JG; Brock, MV. Promoter hypermethylation of resected bronchial margins: a field defect of changes? *Clin Cancer Res*, 2004, 10, 5131-5136.

[64] Harada, K; Toyooka, S; Maitra, A; Maruyama, R; Toyooka, KO; Timmons, CF; Tomlinson, GE; Mastrangelo, D; Hay, RJ; Minna, JD; Gazdar, AF. Aberrant promoter methylation and silencing of the RASSF1A gene in pediatric tumors and cell lines. *Oncogene*, 2002, 21, 4345-4349.

[65] Hasegawa, M; Nelson, HH; Peters, E; Ringstrom, E; Posner, M; Kelsey, KT. Patterns of gene promoter methylation in squamous cell cancer of the head and neck. *Oncogene*, 2002, 21, 4231-4236.

[66] Herman, JG; Baylin, SB. Gene silencing in cancer in association with promoter hypermethylation. *N Engl J Med*, 2003, 349, 2042-2054.

[67] Hesson, L; Dallol, A; Minna, JD; Maher, ER; Latif, F. NORE1A, a homologue of RASSF1A tumour suppressor gene is inactivated in human cancers. *Oncogene*, 2003, 22, 947-954.

[68] Hesson, L; Bieche, I; Krex, D; Criniere, E; Hoang-Xuan, K; Maher, ER; Latif, F. Frequent epigenetic inactivation of RASSF1A and BLU genes located within the critical 3p21.3 region in gliomas. *Oncogene*, 2004, 23, 2408-2419.

[69] Hogg, RP; Honorio, S; Martinez, A; Agathanggelou, A; Dallol, A; Fullwood, P; Weichselbaum, R; Kuo, MJ; Maher, ER; Latif, F. Frequent 3p allele loss and epigenetic inactivation of the RASSF1A tumour suppressor gene from region 3p21.3 in head and neck squamous cell carcinoma. *Eur J Cancer*, 2002, 38, 1585-1592.

[70] Honorio, S; Agathanggelou, A; Schuermann, M; Pankow, W; Viacava, P; Maher, ER; Latif, F. Detection of RASSF1A aberrant promoter hypermethylation in sputum from chronic smokers and ductal carcinoma in situ from breast cancer patients. *Oncogene*, 2003, 22, 147-150.

[71] Honorio, S; Agathanggelou, A; Wernert, N; Rothe, M; Maher, ER; Latif, F. Frequent epigenetic inactivation of the RASSF1A tumour suppressor gene in testicular tumours and distinct methylation profiles of seminoma and nonseminoma testicular germ cell tumours. *Oncogene*, 2003, 22, 461-466.

[72] Hoon, DS; Spugnardi, M; Kuo, C; Huang, SK; Morton, DL; Taback, B. Profiling epigenetic inactivation of tumor suppressor genes in tumors and plasma from cutaneous melanoma patients. *Oncogene*, 2004, 23, 4014-4022.

[73] Horiguchi, K; Tomizawa, Y; Tosaka, M; Ishiuchi, S; Kurihara, H; Mori, M; Saito, N. Epigenetic inactivation of RASSF1A candidate tumor suppressor gene at 3p21.3 in brain tumors. *Oncogene*, 2003, 22, 7862-7865.

[74] House, MG; Guo, M; Efron, DT; Lillemoe, KD; Cameron, JL; Syphard, JE; Hooker, CM; Abraham, SC; Montgomery, EA; Herman, JG; Brock, MV. Tumor suppressor gene hypermethylation as a predictor of gastric stromal tumor behavior. *J Gastrointest Surg*, 2003, 7, 1004-1014; discussion 1014.

[75] House, MG; Herman, JG; Guo, MZ; Hooker, CM; Schulick, RD; Lillemoe, KD; Cameron, JL; Hruban, RH; Maitra, A; Yeo, CJ. Aberrant hypermethylation of tumor suppressor genes in pancreatic endocrine neoplasms. *Ann Surg*, 2003, 238, 423-431; discussion 431-422.

[76] Irimia, M; Fraga, MF; Sanchez-Cespedes, M; Esteller, M. CpG island promoter hypermethylation of the Ras-effector gene NORE1A occurs in the context of a wild-type K-ras in lung cancer. *Oncogene*, 2004, 23, 8695-8699.

[77] Issa, JP. Aging, DNA methylation and cancer. *Crit Rev Oncol Hematol*, 1999, 32, 31-43.

[78] Jackson, JP; Lindroth, AM; Cao, X; Jacobsen, SE. Control of CpNpG DNA methylation by the KRYPTONITE histone H3 methyltransferase. *Nature*, 2002, 416, 556-560.

[79] Jackson, PK. Linking tumor suppression, DNA damage and the anaphase-promoting complex. *Trends Cell Biol*, 2004, 14, 331-334.

[80] Jeddeloh, JA; Stokes, TL; Richards, EJ. Maintenance of genomic methylation requires a SWI2/SNF2-like protein. *Nat Genet*, 1999, 22, 94-97.

[81] Jenuwein, T; Allis, CD. Translating the histone code. *Science*, 2001, 293, 1074-1080.

[82] Jeronimo, C; Henrique, R; Hoque, MO; Mambo, E; Ribeiro, FR; Varzim, G; Oliveira, J; Teixeira, MR; Lopes, C; Sidransky, D. A quantitative promoter methylation profile of prostate cancer. *Clin Cancer Res*, 2004, 10, 8472-8478.

[83] Ji, L; Nishizaki, M; Gao, B; Burbee, D; Kondo, M; Kamibayashi, C; Xu, K; Yen, N; Atkinson, EN; Fang, B; Lerman, MI; Roth, JA; Minna, JD. Expression of several genes in the human chromosome 3p21.3 homozygous deletion region by an adenovirus vector results in tumor suppressor activities in vitro and in vivo. *Cancer Res*, 2002, 62, 2715-2720.

[84] Jones, PA; Wolkowicz, MJ; Rideout, WM, 3rd; Gonzales, FA; Marziasz, CM; Coetzee, GA; Tapscott, SJ. De novo methylation of the MyoD1 CpG island during the establishment of immortal cell lines. *Proc Natl Acad Sci U S A*, 1990, 87, 6117-6121.

[85] Jones, PA; Baylin, SB. The fundamental role of epigenetic events in cancer. *Nat Rev Genet*, 2002, 3, 415-428.

[86] Jones, PL; Veenstra, GJ; Wade, PA; Vermaak, D; Kass, SU; Landsberger, N; Strouboulis, J; Wolffe, AP. Methylated DNA and MeCP2 recruit histone deacetylase to repress transcription. *Nat Genet*, 1998, 19, 187-191.

[87] Kaelin, WG, Jr.; Maher, ER. The VHL tumour-suppressor gene paradigm. *Trends Genet*, 1998, 14, 423-426.

[88] Kang, GH; Lee, S; Kim, WH; Lee, HW; Kim, JC; Rhyu, MG; Ro, JY. Epstein-barr virus-positive gastric carcinoma demonstrates frequent aberrant methylation of multiple genes and constitutes CpG island methylator phenotype-positive gastric carcinoma. *Am J Pathol*, 2002, 160, 787-794.

[89] Kang, GH; Lee, HJ; Hwang, KS; Lee, S; Kim, JH; Kim, JS. Aberrant CpG island hypermethylation of chronic gastritis, in relation to aging, gender, intestinal metaplasia, and chronic inflammation. *Am J Pathol*, 2003, 163, 1551-1556.

[90] Kang, GH; Lee, S; Kim, JS; Jung, HY. Profile of aberrant CpG island methylation along multistep gastric carcinogenesis. *Lab Invest*, 2003, 83, 519-526.

[91] Kang, GH; Lee, S; Lee, HJ; Hwang, KS. Aberrant CpG island hypermethylation of multiple genes in prostate cancer and prostatic intraepithelial neoplasia. *J Pathol*, 2004, 202, 233-240.

[92] Kang, S; Kim, JW; Kang, GH; Park, NH; Song, YS; Kang, SB; Lee, HP. Polymorphism in folate- and methionine-metabolizing enzyme and aberrant CpG island hypermethylation in uterine cervical cancer. *Gynecol Oncol*, 2005, 96, 173-180.

[93] Kawakami, T; Okamoto, K; Kataoka, A; Koizumi, S; Iwaki, H; Sugihara, H; Reeve, AE; Ogawa, O; Okada, Y. Multipoint methylation analysis indicates a distinctive

epigenetic phenotype among testicular germ cell tumors and testicular malignant lymphomas. *Genes Chromosomes Cancer*, 2003, 38, 97-101.

[94] Khokhlatchev, A; Rabizadeh, S; Xavier, R; Nedwidek, M; Chen, T; Zhang, XF; Seed, B; Avruch, J. Identification of a novel Ras-regulated proapoptotic pathway. *Curr Biol*, 2002, 12, 253-265.

[95] Killary, AM; Wolf, ME; Giambernardi, TA; Naylor, SL. Definition of a tumor suppressor locus within human chromosome 3p21-p22. *Proc Natl Acad Sci U S A*, 1992, 89, 10877-10881.

[96] Kim, DH; Kim, JS; Ji, YI; Shim, YM; Kim, H; Han, J; Park, J. Hypermethylation of RASSF1A promoter is associated with the age at starting smoking and a poor prognosis in primary non-small cell lung cancer. *Cancer Res*, 2003, 63, 3743-3746.

[97] Kim, DH; Kim, JS; Park, JH; Lee, SK; Ji, YI; Kwon, YM; Shim, YM; Han, J; Park, J. Relationship of Ras association domain family 1 methylation and K-ras mutation in primary non-small cell lung cancer. *Cancer Res*, 2003, 63, 6206-6211.

[98] Kim, ST; Lim, DS; Canman, CE; Kastan, MB. Substrate specificities and identification of putative substrates of ATM kinase family members. *J Biol Chem*, 1999, 274, 37538-37543.

[99] Kok, K; Naylor, SL; Buys, CH. Deletions of the short arm of chromosome 3 in solid tumors and the search for suppressor genes. *Adv Cancer Res*, 1997, 71, 27-92.

[100] Koul, S; Houldsworth, J; Mansukhani, MM; Donadio, A; McKiernan, JM; Reuter, VE; Bosl, GJ; Chaganti, RS; Murty, VV. Characteristic promoter hypermethylation signatures in male germ cell tumors. *Mol Cancer*, 2002, 1, 8.

[101] Koul, S; McKiernan, JM; Narayan, G; Houldsworth, J; Bacik, J; Dobrzynski, DL; Assaad, AM; Mansukhani, M; Reuter, VE; Bosl, GJ; Chaganti, RS; Murty, VV. Role of promoter hypermethylation in Cisplatin treatment response of male germ cell tumors. *Mol Cancer*, 2004, 3, 16.

[102] Krassenstein, R; Sauter, E; Dulaimi, E; Battagli, C; Ehya, H; Klein-Szanto, A; Cairns, P. Detection of breast cancer in nipple aspirate fluid by CpG island hypermethylation. *Clin Cancer Res*, 2004, 10, 28-32.

[103] Kuroki, T; Trapasso, F; Yendamuri, S; Matsuyama, A; Alder, H; Mori, M; Croce, CM. Promoter hypermethylation of RASSF1A in esophageal squamous cell carcinoma. *Clin Cancer Res*, 2003, 9, 1441-1445.

[104] Kuzmin, I; Gillespie, JW; Protopopov, A; Geil, L; Dreijerink, K; Yang, Y; Vocke, CD; Duh, FM; Zabarovsky, E; Minna, JD; Rhim, JS; Emmert-Buck, MR; Linehan, WM; Lerman, MI. The RASSF1A Tumor Suppressor Gene Is Inactivated in Prostate Tumors and Suppresses Growth of Prostate Carcinoma Cells. *Cancer Res*, 2002, 62, 3498-3502.

[105] Kuzmin, I; Liu, L; Dammann, R; Geil, L; Stanbridge, EJ; Wilczynski, SP; Lerman, MI; Pfeifer, GP. Inactivation of RAS association domain family 1A gene in cervical carcinomas and the role of human papillomavirus infection. *Cancer Res*, 2003, 63, 1888-1893.

[106] Kwong, J; Lo, KW; To, KF; Teo, PM; Johnson, PJ; Huang, DP. Promoter hypermethylation of multiple genes in nasopharyngeal carcinoma. *Clin Cancer Res*, 2002, 8, 131-137.

[107] Lachner, M; O'Carroll, D; Rea, S; Mechtler, K; Jenuwein, T. Methylation of histone H3 lysine 9 creates a binding site for HP1 proteins. *Nature*, 2001, 410, 116-120.

[108] Lee, MG; Kim, HY; Byun, DS; Lee, SJ; Lee, CH; Kim, JI; Chang, SG; Chi, SG. Frequent epigenetic inactivation of rassf1a in human bladder carcinoma. *Cancer Res*, 2001, 61, 6688-6692.

[109] Lee, S; Lee, HJ; Kim, JH; Lee, HS; Jang, JJ; Kang, GH. Aberrant CpG island hypermethylation along multistep hepatocarcinogenesis. *Am J Pathol*, 2003, 163, 1371-1378.

[110] Lee, S; Hwang, KS; Lee, HJ; Kim, JS; Kang, GH. Aberrant CpG island hypermethylation of multiple genes in colorectal neoplasia. *Lab Invest*, 2004, 84, 884-893.

[111] Lehmann, U; Langer, F; Feist, H; Glockner, S; Hasemeier, B; Kreipe, H. Quantitative assessment of promoter hypermethylation during breast cancer development. *Am J Pathol*, 2002, 160, 605-612.

[112] Lehnertz, B; Ueda, Y; Derijck, AA; Braunschweig, U; Perez-Burgos, L; Kubicek, S; Chen, T; Li, E; Jenuwein, T; Peters, AH. Suv39h-mediated histone H3 lysine 9 methylation directs DNA methylation to major satellite repeats at pericentric heterochromatin. *Curr Biol*, 2003, 13, 1192-1200.

[113] Lerman, MI; Minna, JD. The 630-kb lung cancer homozygous deletion region on human chromosome 3p21.3: identification and evaluation of the resident candidate tumor suppressor genes. The International Lung Cancer Chromosome 3p21.3 Tumor Suppressor Gene Consortium. *Cancer Res*, 2000, 60, 6116-6133.

[114] Lewis, CM; Cler, LR; Bu, DW; Zochbauer-Muller, S; Milchgrub, S; Naftalis, EZ; Leitch, AM; Minna, JD; Euhus, DM. Promoter hypermethylation in benign breast epithelium in relation to predicted breast cancer risk. *Clin Cancer Res*, 2005, 11, 166-172.

[115] Li, J; Zhang, Z; Dai, Z; Popkie, AP; Plass, C; Morrison, C; Wang, Y; You, M. RASSF1A promoter methylation and Kras2 mutations in non small cell lung cancer. *Neoplasia*, 2003, 5, 362-366.

[116] Li, J; Wang, F; Protopopov, A; Malyukova, A; Kashuba, V; Minna, JD; Lerman, MI; Klein, G; Zabarovsky, E. Inactivation of RASSF1C during in vivo tumor growth identifies it as a tumor suppressor gene. *Oncogene*, 2004, 23, 5941-5949.

[117] Lim, S; Yang, MH; Park, JH; Nojima, T; Hashimoto, H; Unni, KK; Park, YK. Inactivation of the RASSF1A in osteosarcoma. *Oncol Rep*, 2003, 10, 897-901.

[118] Lindsey, JC; Lusher, ME; Anderton, JA; Bailey, S; Gilbertson, RJ; Pearson, AD; Ellison, DW; Clifford, SC. Identification of tumour-specific epigenetic events in medulloblastoma development by hypermethylation profiling. *Carcinogenesis*, 2004, 25, 661-668.

[119] Liu, L; Amy, V; Liu, G; McKeehan, WL. Novel complex integrating mitochondria and the microtubular cytoskeleton with chromosome remodeling and tumor suppressor RASSF1 deduced by in silico homology analysis, interaction cloning in yeast, and colocalization in cultured cells. *In Vitro Cell Dev Biol Anim*, 2002, 38, 582-594.

[120] Liu, L; Yoon, JH; Dammann, R; Pfeifer, GP. Frequent hypermethylation of the RASSF1A gene in prostate cancer. *Oncogene*, 2002, 21, 6835-6840.

[121] Liu, L; Tommasi, S; Lee, DH; Dammann, R; Pfeifer, GP. Control of microtubule stability by the RASSF1A tumor suppressor. *Oncogene*, 2003, 22, 8125-8136.

[122] Lo, KW; Kwong, J; Hui, AB; Chan, SY; To, KF; Chan, AS; Chow, LS; Teo, PM; Johnson, PJ; Huang, DP. High frequency of promoter hypermethylation of RASSF1A in nasopharyngeal carcinoma. *Cancer Res*, 2001, 61, 3877-3881.

[123] Lusher, ME; Lindsey, JC; Latif, F; Pearson, AD; Ellison, DW; Clifford, SC. Biallelic epigenetic inactivation of the RASSF1A tumor suppressor gene in medulloblastoma development. *Cancer Res*, 2002, 62, 5906-5911.

[124] Marsit, CJ; Kim, DH; Liu, M; Hinds, PW; Wiencke, JK; Nelson, HH; Kelsey, KT. Hypermethylation of RASSF1A and BLU tumor suppressor genes in non-small cell lung cancer: implications for tobacco smoking during adolescence. *Int J Cancer*, 2005, 114, 219-223.

[125] Maruya, S; Issa, JP; Weber, RS; Rosenthal, DI; Haviland, JC; Lotan, R; El-Naggar, AK. Differential methylation status of tumor-associated genes in head and neck squamous carcinoma: incidence and potential implications. *Clin Cancer Res*, 2004, 10, 3825-3830.

[126] Maruyama, R; Toyooka, S; Toyooka, KO; Harada, K; Virmani, AK; Zochbauer-Muller, S; Farinas, AJ; Vakar-Lopez, F; Minna, JD; Sagalowsky, A; Czerniak, B; Gazdar, AF. Aberrant promoter methylation profile of bladder cancer and its relationship to clinicopathological features. *Cancer Res*, 2001, 61, 8659-8663.

[127] Maruyama, R; Toyooka, S; Toyooka, KO; Virmani, AK; Zochbauer-Muller, S; Farinas, AJ; Minna, JD; McConnell, J; Frenkel, EP; Gazdar, AF. Aberrant promoter methylation profile of prostate cancers and its relationship to clinicopathological features. *Clin Cancer Res*, 2002, 8, 514-519.

[128] Maruyama, R; Sugio, K; Yoshino, I; Maehara, Y; Gazdar, AF. Hypermethylation of FHIT as a prognostic marker in nonsmall cell lung carcinoma. *Cancer*, 2004, 100, 1472-1477.

[129] Mathe, E. RASSF1A, the new guardian of mitosis. *Nat Genet*, 2004, 36, 117-118.

[130] Mayo, MW; Wang, CY; Cogswell, PC; Rogers-Graham, KS; Lowe, SW; Der, CJ; Baldwin, AS, Jr. Requirement of NF-kappaB activation to suppress p53-independent apoptosis induced by oncogenic Ras. *Science*, 1997, 278, 1812-1815.

[131] Mehrotra, J; Vali, M; McVeigh, M; Kominsky, SL; Fackler, MJ; Lahti-Domenici, J; Polyak, K; Sacchi, N; Garrett-Mayer, E; Argani, P; Sukumar, S. Very high frequency of hypermethylated genes in breast cancer metastasis to the bone, brain, and lung. *Clin Cancer Res*, 2004, 10, 3104-3109.

[132] Morrissey, C; Martinez, A; Zatyka, M; Agathanggelou, A; Honorio, S; Astuti, D; Morgan, NV; Moch, H; Richards, FM; Kishida, T; Yao, M; Schraml, P; Latif, F; Maher, ER. Epigenetic inactivation of the RASSF1A 3p21.3 tumor suppressor gene in both clear cell and papillary renal cell carcinoma. *Cancer Res*, 2001, 61, 7277-7281.

[133] Muller, HM; Widschwendter, A; Fiegl, H; Ivarsson, L; Goebel, G; Perkmann, E; Marth, C; Widschwendter, M. DNA methylation in serum of breast cancer patients: an independent prognostic marker. *Cancer Res*, 2003, 63, 7641-7645.

[134] Murray, PG; Qiu, GH; Fu, L; Waites, ER; Srivastava, G; Heys, D; Agathanggelou, A; Latif, F; Grundy, RG; Mann, JR; Starczynski, J; Crocker, J; Parkes, SE; Ambinder, RF;

Young, LS; Tao, Q. Frequent epigenetic inactivation of the RASSF1A tumor suppressor gene in Hodgkin's lymphoma. *Oncogene*, 2004, 23, 1326-1331.

[135] Nakayama, J; Rice, JC; Strahl, BD; Allis, CD; Grewal, SI. Role of histone H3 lysine 9 methylation in epigenetic control of heterochromatin assembly. *Science*, 2001, 292, 110-113.

[136] Nan, X; Ng, HH; Johnson, CA; Laherty, CD; Turner, BM; Eisenman, RN; Bird, A. Transcriptional repression by the methyl-CpG-binding protein MeCP2 involves a histone deacetylase complex. *Nature*, 1998, 393, 386-389.

[137] Newton, AC. Protein kinase C. Seeing two domains. *Curr Biol*, 1995, 5, 973-976.

[138] Ng, HH; Zhang, Y; Hendrich, B; Johnson, CA; Turner, BM; Erdjument-Bromage, H; Tempst, P; Reinberg, D; Bird, A. MBD2 is a transcriptional repressor belonging to the MeCP1 histone deacetylase complex. *Nat Genet*, 1999, 23, 58-61.

[139] Ng, MH; Lau, KM; Wong, WS; To, KW; Cheng, SH; Tsang, KS; Chan, NP; Kho, BC; Lo, KW; Tong, JH; Lam, CW; Chan, JC. Alterations of RAS signalling in Chinese multiple myeloma patients: absent BRAF and rare RAS mutations, but frequent inactivation of RASSF1A by transcriptional silencing or expression of a non-functional variant transcript. *Br J Haematol*, 2003, 123, 637-645.

[140] Ortiz-Vega, S; Khokhlatchev, A; Nedwidek, M; Zhang, XF; Dammann, R; Pfeifer, GP; Avruch, J. The putative tumor suppressor RASSF1A homodimerizes and heterodimerizes with the Ras-GTP binding protein Nore1. *Oncogene*, 2002, 21, 1381-1390.

[141] Ponting, CP; Benjamin, DR. A novel family of Ras-binding domains. *Trends Biochem Sci*, 1996, 21, 422-425.

[142] Praskova, M; Khoklatchev, A; Ortiz-Vega, S; Avruch, J. Regulation of the MST1 kinase by autophosphorylation, by the growth inhibitory proteins, RASSF1 and NORE1, and by Ras. *Biochem J*, 2004, 381, 453-462.

[143] Rabizadeh, S; Xavier, RJ; Ishiguro, K; Bernabeortiz, J; Lopez-Ilasaca, M; Khokhlatchev, A; Mollahan, P; Pfeifer, GP; Avruch, J; Seed, B. The scaffold protein CNK1 interacts with the tumor suppressor RASSF1A and augments RASSF1A-induced cell death. *J Biol Chem*, 2004, 279, 29247-29254.

[144] Ramirez, JL; Sarries, C; de Castro, PL; Roig, B; Queralt, C; Escuin, D; de Aguirre, I; Sanchez, JM; Manzano, JL; Margeli, M; Sanchez, JJ; Astudillo, J; Taron, M; Rosell, R. Methylation patterns and K-ras mutations in tumor and paired serum of resected non-small-cell lung cancer patients. *Cancer Lett*, 2003, 193, 207-216.

[145] Ramirez, JL; Taron, M; Balana, C; Sarries, C; Mendez, P; de Aguirre, I; Nunez, L; Roig, B; Queralt, C; Botia, M; Rosell, R. Serum DNA as a tool for cancer patient management. *Rocz Akad Med Bialymst*, 2003, 48, 34-41.

[146] Rathi, A; Virmani, AK; Schorge, JO; Elias, KJ; Maruyama, R; Minna, JD; Mok, SC; Girard, L; Fishman, DA; Gazdar, AF. Methylation profiles of sporadic ovarian tumors and nonmalignant ovaries from high-risk women. *Clin Cancer Res*, 2002, 8, 3324-3331.

[147] Reifenberger, J; Knobbe, CB; Sterzinger, AA; Blaschke, B; Schulte, KW; Ruzicka, T; Reifenberger, G. Frequent alterations of Ras signaling pathway genes in sporadic malignant melanomas. *Int J Cancer*, 2004, 109, 377-384.

[148] Robertson, KD; Ait-Si-Ali, S; Yokochi, T; Wade, PA; Jones, PL; Wolffe, AP. DNMT1 forms a complex with Rb, E2F1 and HDAC1 and represses transcription from E2F-responsive promoters. *Nat Genet*, 2000, 25, 338-342.

[149] Rong, R; Jin, W; Zhang, J; Saeed Sheikh, M; Huang, Y. Tumor suppressor RASSF1A is a microtubule-binding protein that stabilizes microtubules and induces G(2)/M arrest. *Oncogene*, 2004.

[150] Rong, R; Jin, W; Zhang, J; Sheikh, MS; Huang, Y. Tumor suppressor RASSF1A is a microtubule-binding protein that stabilizes microtubules and induces G2/M arrest. *Oncogene*, 2004, 23, 8216-8230.

[151] Rountree, MR; Bachman, KE; Baylin, SB. DNMT1 binds HDAC2 and a new co-repressor, DMAP1, to form a complex at replication foci. *Nat Genet*, 2000, 25, 269-277.

[152] Sakamoto, N; Terai, T; Ajioka, Y; Abe, S; Kobayasi, O; Hirai, S; Hino, O; Watanabe, H; Sato, N; Shimoda, T; Fujii, H. Frequent hypermethylation of RASSF1A in early flat-type colorectal tumors. *Oncogene*, 2004, 23, 8900-8907.

[153] Sanchez-Cespedes, M; Esteller, M; Wu, L; Nawroz-Danish, H; Yoo, GH; Koch, WM; Jen, J; Herman, JG; Sidransky, D. Gene promoter hypermethylation in tumors and serum of head and neck cancer patients. *Cancer Res*, 2000, 60, 892-895.

[154] Schagdarsurengin, U; Gimm, O; Hoang-Vu, C; Dralle, H; Pfeifer, GP; Dammann, R. Frequent epigenetic silencing of the CpG island promoter of RASSF1A in thyroid carcinoma. *Cancer Res*, 2002, 62, 3698-3701.

[155] Schagdarsurengin, U; Wilkens, L; Steinemann, D; Flemming, P; Kreipe, HH; Pfeifer, GP; Schlegelberger, B; Dammann, R. Frequent epigenetic inactivation of the RASSF1A gene in hepatocellular carcinoma. *Oncogene*, 2003, 22, 1866-1871.

[156] Seidel, C; Bartel, F; Rastetter, M; Bluemke, K; Wurl, P; Taubert, H; Dammann, R. Alterations of cancer-related genes in soft tissue sarcomas: Hypermethylation of RASSF1A is frequently detected in leiomyosarcoma and associated with poor prognosis in sarcoma. *Int J Cancer*, 2005, 114, 442-447.

[157] Seidl, S; Ackermann, J; Kaufmann, H; Keck, A; Nosslinger, T; Zielinski, CC; Drach, J; Zochbauer-Muller, S. DNA-methylation analysis identifies the E-cadherin gene as a potential marker of disease progression in patients with monoclonal gammopathies. *Cancer*, 2004, 100, 2598-2606.

[158] Sekido, Y; Ahmadian, M; Wistuba, II; Latif, F; Bader, S; Wei, MH; Duh, FM; Gazdar, AF; Lerman, MI; Minna, JD. Cloning of a breast cancer homozygous deletion junction narrows the region of search for a 3p21.3 tumor suppressor gene. *Oncogene*, 1998, 16, 3151-3157.

[159] Serrano, M; Lin, AW; McCurrach, ME; Beach, D; Lowe, SW. Oncogenic ras provokes premature cell senescence associated with accumulation of p53 and p16INK4a. *Cell*, 1997, 88, 593-602.

[160] Shao, J; Sheng, H; DuBois, RN; Beauchamp, RD. Oncogenic Ras-mediated cell growth arrest and apoptosis are associated with increased ubiquitin-dependent cyclin D1 degradation. *J Biol Chem*, 2000, 275, 22916-22921.

[161] Shivakumar, L; Minna, J; Sakamaki, T; Pestell, R; White, MA. The RASSF1A tumor suppressor blocks cell cycle progression and inhibits cyclin D1 accumulation. *Mol Cell Biol*, 2002, 22, 4309-4318.

[162] Singal, R; Ferdinand, L; Reis, IM; Schlesselman, JJ. Methylation of multiple genes in prostate cancer and the relationship with clinicopathological features of disease. *Oncol Rep*, 2004, 12, 631-637.

[163] Smiraglia, DJ; Rush, LJ; Fruhwald, MC; Dai, Z; Held, WA; Costello, JF; Lang, JC; Eng, C; Li, B; Wright, FA; Caligiuri, MA; Plass, C. Excessive CpG island hypermethylation in cancer cell lines versus primary human malignancies. *Hum Mol Genet*, 2001, 10, 1413-1419.

[164] Smith, AJ; Xian, J; Richardson, M; Johnstone, KA; Rabbitts, PH. Cre-loxP chromosome engineering of a targeted deletion in the mouse corresponding to the 3p21.3 region of homozygous loss in human tumours. *Oncogene*, 2002, 21, 4521-4529.

[165] Song, MS; Song, SJ; Ayad, NG; Chang, JS; Lee, JH; Hong, HK; Lee, H; Choi, N; Kim, J; Kim, H; Kim, JW; Choi, EJ; Kirschner, MW; Lim, DS. The tumour suppressor RASSF1A regulates mitosis by inhibiting the APC-Cdc20 complex. *Nat Cell Biol*, 2004, 6, 129-137.

[166] Song, MS; Chang, JS; Song, SJ; Yang, TH; Lee, H; Lim, DS. The centrosomal protein RAS association domain family protein 1A (RASSF1A)-binding protein 1 regulates mitotic progression by recruiting RASSF1A to spindle poles. *J Biol Chem*, 2005, 280, 3920-3927.

[167] Sozzi, G; Veronese, ML; Negrini, M; Baffa, R; Cotticelli, MG; Inoue, H; Tornielli, S; Pilotti, S; De Gregorio, L; Pastorino, U; Pierotti, MA; Ohta, M; Huebner, K; Croce, CM. The FHIT gene 3p14.2 is abnormal in lung cancer. *Cell*, 1996, 85, 17-26.

[168] Spugnardi, M; Tommasi, S; Dammann, R; Pfeifer, GP; Hoon, DS. Epigenetic inactivation of RAS association domain family protein 1 (RASSF1A) in malignant cutaneous melanoma. *Cancer Res*, 2003, 63, 1639-1643.

[169] Takahashi, T; Shivapurkar, N; Riquelme, E; Shigematsu, H; Reddy, J; Suzuki, M; Miyajima, K; Zhou, X; Bekele, BN; Gazdar, AF; Wistuba, II. Aberrant promoter hypermethylation of multiple genes in gallbladder carcinoma and chronic cholecystitis. *Clin Cancer Res*, 2004, 10, 6126-6133.

[170] Tamaru, H; Selker, EU. A histone H3 methyltransferase controls DNA methylation in Neurospora crassa. *Nature*, 2001, 414, 277-283.

[171] To, KF; Leung, WK; Lee, TL; Yu, J; Tong, JH; Chan, MW; Ng, EK; Chung, SC; Sung, JJ. Promoter hypermethylation of tumor-related genes in gastric intestinal metaplasia of patients with and without gastric cancer. *Int J Cancer*, 2002, 102, 623-628.

[172] Todd, S; Franklin, WA; Varella-Garcia, M; Kennedy, T; Hilliker, CE, Jr.; Hahner, L; Anderson, M; Wiest, JS; Drabkin, HA; Gemmill, RM. Homozygous deletions of human chromosome 3p in lung tumors. *Cancer Res*, 1997, 57, 1344-1352.

[173] Tokinaga, K; Okuda, H; Nomura, A; Ashida, S; Furihata, M; Shuin, T. Hypermethylation of the RASSF1A tumor suppressor gene in Japanese clear cell renal cell carcinoma. *Oncol Rep*, 2004, 12, 805-810.

[174] Tomizawa, Y; Kohno, T; Kondo, H; Otsuka, A; Nishioka, M; Niki, T; Yamada, T; Maeshima, A; Yoshimura, K; Saito, R; Minna, JD; Yokota, J. Clinicopathological

Significance of Epigenetic Inactivation of RASSF1A at 3p21.3 in Stage I Lung Adenocarcinoma. *Clin Cancer Res*, 2002, 8, 2362-2368.

[175] Tomizawa, Y; Iijima, H; Nomoto, T; Iwasaki, Y; Otani, Y; Tsuchiya, S; Saito, R; Dobashi, K; Nakajima, T; Mori, M. Clinicopathological significance of aberrant methylation of RARbeta2 at 3p24, RASSF1A at 3p21.3, and FHIT at 3p14.2 in patients with non-small cell lung cancer. *Lung Cancer*, 2004, 46, 305-312.

[176] Tommasi, S; Dammann, R; Jin, SG; Zhang Xf, XF; Avruch, J; Pfeifer, GP. RASSF3 and NORE1: identification and cloning of two human homologues of the putative tumor suppressor gene RASSF1. *Oncogene*, 2002, 21, 2713-2720.

[177] Tommasi, S; Dammann, R; Zhang, Z; Wang, Y; Liu, L; Tsark, WM; Wilczynski, SP; Li, J; You, M; Pfeifer, GP. Tumor susceptibility of Rassf1a knockout mice. *Cancer Res*, 2005, 65, 92-98.

[178] Tong, JH; Tsang, RK; Lo, KW; Woo, JK; Kwong, J; Chan, MW; Chang, AR; van Hasselt, CA; Huang, DP; To, KF. Quantitative Epstein-Barr virus DNA analysis and detection of gene promoter hypermethylation in nasopharyngeal (NP) brushing samples from patients with NP carcinoma. *Clin Cancer Res*, 2002, 8, 2612-2619.

[179] Topaloglu, O; Hoque, MO; Tokumaru, Y; Lee, J; Ratovitski, E; Sidransky, D; Moon, CS. Detection of promoter hypermethylation of multiple genes in the tumor and bronchoalveolar lavage of patients with lung cancer. *Clin Cancer Res*, 2004, 10, 2284-2288.

[180] Toyooka, S; Pass, HI; Shivapurkar, N; Fukuyama, Y; Maruyama, R; Toyooka, KO; Gilcrease, M; Farinas, A; Minna, JD; Gazdar, AF. Aberrant methylation and simian virus 40 tag sequences in malignant mesothelioma. *Cancer Res*, 2001, 61, 5727-5730.

[181] Toyooka, S; Toyooka, KO; Maruyama, R; Virmani, AK; Girard, L; Miyajima, K; Harada, K; Ariyoshi, Y; Takahashi, T; Sugio, K; Brambilla, E; Gilcrease, M; Minna, JD; Gazdar, AF. DNA methylation profiles of lung tumors. *Mol Cancer Ther*, 2001, 1, 61-67.

[182] Toyooka, S; Carbone, M; Toyooka, KO; Bocchetta, M; Shivapurkar, N; Minna, JD; Gazdar, AF. Progressive aberrant methylation of the RASSF1A gene in simian virus 40 infected human mesothelial cells. *Oncogene*, 2002, 21, 4340-4344.

[183] Toyooka, S; Maruyama, R; Toyooka, KO; McLerran, D; Feng, Z; Fukuyama, Y; Virmani, AK; Zochbauer-Muller, S; Tsukuda, K; Sugio, K; Shimizu, N; Shimizu, K; Lee, H; Chen, CY; Fong, KM; Gilcrease, M; Roth, JA; Minna, JD; Gazdar, AF. Smoke exposure, histologic type and geography-related differences in the methylation profiles of non-small cell lung cancer. *Int J Cancer*, 2003, 103, 153-160.

[184] Toyooka, S; Suzuki, M; Tsuda, T; Toyooka, KO; Maruyama, R; Tsukuda, K; Fukuyama, Y; Iizasa, T; Fujisawa, T; Shimizu, N; Minna, JD; Gazdar, AF. Dose effect of smoking on aberrant methylation in non-small cell lung cancers. *Int J Cancer*, 2004, 110, 462-464.

[185] Tozawa, T; Tamura, G; Honda, T; Nawata, S; Kimura, W; Makino, N; Kawata, S; Sugai, T; Suto, T; Motoyama, T. Promoter hypermethylation of DAP-kinase is associated with poor survival in primary biliary tract carcinoma patients. *Cancer Sci*, 2004, 95, 736-740.

[186] Tsou, JA; Hagen, JA; Carpenter, CL; Laird-Offringa, IA. DNA methylation analysis: a powerful new tool for lung cancer diagnosis. *Oncogene*, 2002, 21, 5450-5461.

[187] Turker, MS. Gene silencing in mammalian cells and the spread of DNA methylation. *Oncogene*, 2002, 21, 5388-5393.

[188] van Engeland, M; Roemen, GM; Brink, M; Pachen, MM; Weijenberg, MP; de Bruine, AP; Arends, JW; van den Brandt, PA; de Goeij, AF; Herman, JG. K-ras mutations and RASSF1A promoter methylation in colorectal cancer. *Oncogene*, 2002, 21, 3792-3795.

[189] van Engeland, M; Weijenberg, MP; Roemen, GM; Brink, M; de Bruine, AP; Goldbohm, RA; van den Brandt, PA; Baylin, SB; de Goeij, AF; Herman, JG. Effects of dietary folate and alcohol intake on promoter methylation in sporadic colorectal cancer: the Netherlands cohort study on diet and cancer. *Cancer Res*, 2003, 63, 3133-3137.

[190] Vavvas, D; Li, X; Avruch, J; Zhang, XF. Identification of Nore1 as a potential Ras effector. *J Biol Chem*, 1998, 273, 5439-5442.

[191] Vos, MD; Ellis, CA; Bell, A; Birrer, MJ; Clark, GJ. Ras uses the novel tumor suppressor RASSF1 as an effector to mediate apoptosis. *J Biol Chem*, 2000, 275, 35669-35672.

[192] Vos, MD; Martinez, A; Ellis, CA; Vallecorsa, T; Clark, GJ. The pro-apoptotic Ras effector Nore1 may serve as a Ras-regulated tumor suppressor in the lung. *J Biol Chem*, 2003, 278, 21938-21943.

[193] Vos, MD; Martinez, A; Elam, C; Dallol, A; Taylor, BJ; Latif, F; Clark, GJ. A role for the RASSF1A tumor suppressor in the regulation of tubulin polymerization and genomic stability. *Cancer Res*, 2004, 64, 4244-4250.

[194] Wagner, KJ; Cooper, WN; Grundy, RG; Caldwell, G; Jones, C; Wadey, RB; Morton, D; Schofield, PN; Reik, W; Latif, F; Maher, ER. Frequent RASSF1A tumour suppressor gene promoter methylation in Wilms' tumour and colorectal cancer. *Oncogene*, 2002, 21, 7277-7282.

[195] Waki, T; Tamura, G; Sato, M; Motoyama, T. Age-related methylation of tumor suppressor and tumor-related genes: an analysis of autopsy samples. *Oncogene*, 2003, 22, 4128-4133.

[196] Wang, J; Lee, JJ; Wang, L; Liu, DD; Lu, C; Fan, YH; Hong, WK; Mao, L. Value of p16INK4a and RASSF1A promoter hypermethylation in prognosis of patients with resectable non-small cell lung cancer. *Clin Cancer Res*, 2004, 10, 6119-6125.

[197] Wei, MH; Latif, F; Bader, S; Kashuba, V; Chen, JY; Duh, FM; Sekido, Y; Lee, CC; Geil, L; Kuzmin, I; Zabarovsky, E; Klein, G; Zbar, B; Minna, JD; Lerman, MI. Construction of a 600-kilobase cosmid clone contig and generation of a transcriptional map surrounding the lung cancer tumor suppressor gene (TSG) locus on human chromosome 3p21.3: progress toward the isolation of a lung cancer TSG. *Cancer Res*, 1996, 56, 1487-1492.

[198] Wistuba, II; Behrens, C; Virmani, AK; Mele, G; Milchgrub, S; Girard, L; Fondon, JW, 3rd; Garner, HR; McKay, B; Latif, F; Lerman, MI; Lam, S; Gazdar, AF; Minna, JD. High resolution chromosome 3p allelotyping of human lung cancer and preneoplastic/preinvasive bronchial epithelium reveals multiple, discontinuous sites of 3p allele loss and three regions of frequent breakpoints. *Cancer Res*, 2000, 60, 1949-1960.

[199] Wong, IH; Lo, YM; Zhang, J; Liew, CT; Ng, MH; Wong, N; Lai, PB; Lau, WY; Hjelm, NM; Johnson, PJ. Detection of aberrant p16 methylation in the plasma and serum of liver cancer patients. *Cancer Res*, 1999, 59, 71-73.

[200] Wong, IH; Chan, J; Wong, J; Tam, PK. Ubiquitous aberrant RASSF1A promoter methylation in childhood neoplasia. *Clin Cancer Res*, 2004, 10, 994-1002.

[201] Wong, N; Li, L; Tsang, K; Lai, PB; To, KF; Johnson, PJ. Frequent loss of chromosome 3p and hypermethylation of RASSF1A in cholangiocarcinoma. *J Hepatol*, 2002, 37, 633-639.

[202] Wong, TS; Tang, KC; Kwong, DL; Sham, JS; Wei, WI; Kwong, YL; Yuen, AP. Differential gene methylation in undifferentiated nasopharyngeal carcinoma. *Int J Oncol*, 2003, 22, 869-874.

[203] Wong, TS; Kwong, DL; Sham, JS; Wei, WI; Kwong, YL; Yuen, AP. Quantitative plasma hypermethylated DNA markers of undifferentiated nasopharyngeal carcinoma. *Clin Cancer Res*, 2004, 10, 2401-2406.

[204] Woodson, K; Gillespie, J; Hanson, J; Emmert-Buck, M; Phillips, JM; Linehan, WM; Tangrea, JA. Heterogeneous gene methylation patterns among pre-invasive and cancerous lesions of the prostate: a histopathologic study of whole mount prostate specimens. *Prostate*, 2004, 60, 25-31.

[205] Woodson, K; Hanson, J; Tangrea, J. A survey of gene-specific methylation in human prostate cancer among black and white men. *Cancer Lett*, 2004, 205, 181-188.

[206] Xian, J; Clark, KJ; Fordham, R; Pannell, R; Rabbitts, TH; Rabbitts, PH. Inadequate lung development and bronchial hyperplasia in mice with a targeted deletion in the Dutt1/Robo1 gene. *Proc Natl Acad Sci U S A*, 2001, 98, 15062-15066.

[207] Xing, M; Cohen, Y; Mambo, E; Tallini, G; Udelsman, R; Ladenson, PW; Sidransky, D. Early occurrence of RASSF1A hypermethylation and its mutual exclusion with BRAF mutation in thyroid tumorigenesis. *Cancer Res*, 2004, 64, 1664-1668.

[208] Xu, XL; Yu, J; Zhang, HY; Sun, MH; Gu, J; Du, X; Shi, DR; Wang, P; Yang, ZH; Zhu, JD. Methylation profile of the promoter CpG islands of 31 genes that may contribute to colorectal carcinogenesis. *World J Gastroenterol*, 2004, 10, 3441-3454.

[209] Yamakawa, K; Takahashi, T; Horio, Y; Murata, Y; Takahashi, E; Hibi, K; Yokoyama, S; Ueda, R; Nakamura, Y. Frequent homozygous deletions in lung cancer cell lines detected by a DNA marker located at 3p21.3-p22. *Oncogene*, 1993, 8, 327-330.

[210] Yanagawa, N; Tamura, G; Oizumi, H; Takahashi, N; Shimazaki, Y; Motoyama, T. Promoter hypermethylation of tumor suppressor and tumor-related genes in non-small cell lung cancers. *Cancer Sci*, 2003, 94, 589-592.

[211] Yang, B; House, MG; Guo, M; Herman, JG; Clark, DP. Promoter methylation profiles of tumor suppressor genes in intrahepatic and extrahepatic cholangiocarcinoma. *Mod Pathol*, 2005, 18, 412-420.

[212] Yang, Q; Zage, P; Kagan, D; Tian, Y; Seshadri, R; Salwen, HR; Liu, S; Chlenski, A; Cohn, SL. Association of epigenetic inactivation of RASSF1A with poor outcome in human neuroblastoma. *Clin Cancer Res*, 2004, 10, 8493-8500.

[213] Yegnasubramanian, S; Kowalski, J; Gonzalgo, ML; Zahurak, M; Piantadosi, S; Walsh, PC; Bova, GS; De Marzo, AM; Isaacs, WB; Nelson, WG. Hypermethylation of CpG

islands in primary and metastatic human prostate cancer. *Cancer Res*, 2004, 64, 1975-1986.

[214] Yoon, JH; Dammann, R; Pfeifer, GP. Hypermethylation of the CpG island of the RASSF1A gene in ovarian and renal cell carcinomas. *Int J Cancer*, 2001, 94, 212-217.

[215] Yu, J; Zhang, H; Gu, J; Lin, S; Li, J; Lu, W; Wang, Y; Zhu, J. Methylation profiles of thirty four promoter-CpG islands and concordant methylation behaviours of sixteen genes that may contribute to carcinogenesis of astrocytoma. *BMC Cancer*, 2004, 4, 65.

[216] Yu, MY; Tong, JH; Chan, PK; Lee, TL; Chan, MW; Chan, AW; Lo, KW; To, KF. Hypermethylation of the tumor suppressor gene RASSFIA and frequent concomitant loss of heterozygosity at 3p21 in cervical cancers. *Int J Cancer*, 2003, 105, 204-209.

[217] Zhong, S; Yeo, W; Tang, MW; Wong, N; Lai, PB; Johnson, PJ. Intensive hypermethylation of the CpG island of Ras association domain family 1A in hepatitis B virus-associated hepatocellular carcinomas. *Clin Cancer Res*, 2003, 9, 3376-3382.

[218] Zochbauer-Muller, S; Lam, S; Toyooka, S; Virmani, AK; Toyooka, KO; Seidl, S; Minna, JD; Gazdar, AF. Aberrant methylation of multiple genes in the upper aerodigestive tract epithelium of heavy smokers. *Int J Cancer*, 2003, 107, 612-616.

In: Gene Silencing: New Research
Editor: Grace W. Redberry, pp. 121-141
ISBN 1-59454-832-3

Chapter IV

Human DNA cis-Elements Protecting the VHL Promoter and CpG Island from Methylation are Functional in the Mouse

Igor Kuzmin[1,*] *Laura Geil*[1], *Laura Schmidt*[1], *Fuh-Mei Duh*[1], *Lionel Feigenbaum*[2] *and Michael I. Lerman*[3]
Basic Research Program[1], Transgenic Mouse Model, SAIC-Frederick, Inc.[2], Laboratory of Immunobiology[3], NCI Frederick, Frederick, Maryland

Abstract

A prerequisite for understanding the functions and mechanisms of DNA methylation in mammals is knowledge of its distribution in the genome. However, the detailed long-range local methylation profiles are not available. We used bisulfite sequencing to obtain high-resolution methylation maps of the transcriptionally active mouse and human VHL loci. The VHL CpG island, including the promoter and the first exon, is not methylated in normal tissues, but hypermethylated in some sporadic renal carcinomas. Our maps cover 17 kb of human and 6.7 kb of mouse genomic DNA and include 434 and 151 CpG dinucleotides respectively. Despite the divergence of genomic sequences, similar methylation patterns were maintained in mouse and human cells. Both loci contained 1.5 kb methylation-protected regions, including a 1 kb GC-rich island and 0.5 kb of the adjacent part of intron 1. All other sequences, unique and repetitive alike, were heavily methylated. We then developed transgenic mice carrying the human VHL CpG island and the adjacent sequences to begin studying the nature of mouse and human DNA *cis* elements protecting these CpG islands from methylation. In 3 of 5 mouse strains

* Corresponding author. Mailing address: Laboratory of Immunobiology, bldg. 560, rm. 12-34, SAIC-Frederick, Inc., NCI Frederick, Frederick, MD 21702. Phone: (301) 846-7320. Fax: (301) 846-6145. E-mail: kuzmin@mail.ncifcrf.gov

originating from independent founder animals, methylation patterns of human transgenes resembled those of normal human cells indicating that the essential *cis* DNA protecting elements were present within the transgene and functional in the mouse environment. Developing mice, which form and maintain normal methylation patterns on human DNA templates, make possible systematic dissection of elements relevant for formation and maintenance of methylation patterns in tumor and normal cells.

Introduction

Approximately 70 to 80 % of cytosines within CpG dinucleotides are non-randomly methylated in the genomic DNA of normal differentiated mammalian cells [2]. Formation and maintenance of these specific DNA methylation patterns are dependent upon local chromatin organization. These patterns play an important role in carcinogenesis and normal development, effecting gene expression [17]. The majority of CpG dinucleotides within functional gene promoters and other regulatory elements are not methylated. Many of these regulatory regions are GC-rich and have a higher density of CpG dinucleotides than the genomic average. These GC-rich regions are called CpG islands. Aberrant hypermethylation of the CpG islands causes silencing of critical genes, in some cases, leading to cancer and developmental abnormalities. [4, 29]. In contrast, short interspersed repetitive elements within the bodies of the genes and intergenic regions are often hypermethylated.

The mechanisms which induce and maintain these normal or aberrant methylation patterns, remain largely unknown and include two alternatives [2]. According to the first model, suppressive methylation is limited predominantly to interspersed repetitive elements. In the second model, methylation may be able to spread to adjacent "unique" genomic sequences. In this case, the critical elements controlling gene expression would have to be protected from methylation spreading by locus-specific mechanisms. For a limited number of loci, such protecting mechanisms have been suggested [24, 31]. To determine which of these two alternative methylation mechanisms is functional within the VHL locus, the assembly of a high-resolution long-range continuous methylation map is first necessary. Such a map should include CpG islands, interspersed repeats and "unique" genomic sequences. However, available methylation maps are sketchy and usually provide information for very short stretches of DNA sequence [25, 32], limited by the size of PCR fragments obtained from fragmented bisulfite-treated DNA [8].

Additionally, the experimental systems suitable to study formation of local methylation patterns are not readily available and problematic. Transfection of naked unmethylated DNA into somatic mammalian cells does not cause re-establishment of proper methylation of the transgene even for the largest unmethylated DNA fragments. In somatic cells *de novo* methylation (if any) proceeds at an extremely slow rate [18,20], making any analysis of locus-specific methylation protection difficult. Since most *de novo* methylation occurs naturally in embryos, shortly before implantation [27], the normal transgene's *de novo* methylation is expected to take place during development. However, methylation analysis of a transgene in the cells of the same species may be complicated by the presence of identical endogenous sequences. Mouse embryos, stem cells, teratocarcinomas and other hosts possessing efficient *de novo* methylation machinery may be suitable to study methylation protection of human CpG islands, although, the cross-species interchangeability of *cis* DNA elements and *trans*-acting protein factors requires verification. To study the feasibility of such heterologous systems for investigation of DNA methylation protection mechanisms, we used the VHL locus as a model.

The human VHL gene [19] spans a 12 kb region and includes three exons. The first exon and promoter are within a CpG island [16]. The gene was shown to be silenced by aberrant methylation of its promoter in a large percentage of VHL-specific tumors [10]. In VHL-expressing cells the island is unmethylated. Some CpG methylation was found in two adjacent Alu repeats [6,18], implying the existence of protecting elements on the borders of the island. Concurrently, other parts of intron 1 remained unmethylated [18] raising the possibility that low levels of methylation represent a common feature of the entire VHL locus at 3p25-p26 and that no special methylation protection mechanisms exist for the island itself. To resolve these alternatives, we extended our high-resolution methylation map to the entire human VHL gene (17 kb) and localized the borders of the region protected from DNA methylation. In the mouse VHL locus, we sequenced 9 kb of genomic DNA including all three exons and the adjacent genomic sequences. Using these data we built a high-resolution methylation map, which includes 6.7 kb of genomic DNA around the CpG island. Finally to study the methylation-protecting machinery, we analyzed whether the human pattern of VHL locus methylation could be preserved in the mouse background, using trangenic mice.

Materials and Methods

Sequence Assembly of the Mouse VHL Locus

A unique Not I site is located within the mouse VHL exon 1. P1 clone 1960 [26] was obtained from Genome Systems, Inc. using mouse VHL cDNA as a probe. Nsi I-Not I digestion of P1 clone 1960 produced a 2.8 kb fragment detected by hybridization with the VHL exon 1 probe. This fragment containing the region upstream of the VHL promoter (Nsi I-EcoRI fragment) was subcloned into pBluescript II SK (Stratagene, La Jolla, CA) and sequenced. An adjacent downstream 3.5 kb EcoRI fragment, which includes the putative promoter, exon 1 and a part of intron 1, was subcloned and sequenced separately.

129VJ mouse genomic library in Lambda FixII vector (Stratagene) was screened with the 1.3 kb full length M. musculus VHL cDNA. A phage clone was selected containing intron 1, exons 2 and 3, and the 3'UTR of the VHL gene. The 15 kb insert was released by Not I, subcloned into pBluescript II SK vector and designated VHL clone 9. A part of the clone 9 insert containing the remaining VHL exons and introns was sequenced.

The assembled sequence of the VHL gene, including putative promoter, exons, introns and flanking genomic sequences, was deposited in GenBank under accession number AF513984.

Bisulfite Sequencing

Methylation analyses of human and mouse VHL loci were performed using DNA isolated from human UMRC6 cells [7]. The cells were kindly provided by H.B. Grossman. Mouse VHL bisulfite sequencing was performed using DNA from NIH-3T3 cells [14], purchased from ATCC (Manassas, VA). Primers for amplification of bisulfite-treated DNA are listed in the Appendix, Table A1 and A2. The first and nested PCR reactions were performed at 94°C, 30 sec; 58°C, 30 sec; 72°C, 2 min; for 35 cycles. PCR fragments were cloned into plasmid pCR 2.1 TOPO (Invitrigen, Carlsbad, CA) and sequenced using a 373 DNA Stretch automatic sequencer (PE-Applied Biosystems, Foster City, CA).

Preparation of Transgenic Mice

A NheI/SacI 9.5 kb human genomic fragment including the VHL CpG island (with VHL promoter and exon 1), 2.5 kb of the promoter upstream sequence, intron 1, exon 2 and a part (2 kb) of intron 2 was subcloned into plasmid pBluescript II SK (Stratagene, La Jolla, CA). For microinjection, the plasmid was digested with KpnI and SacI and the insert purified on 10 to 40 % sucrose gradient by ultracentrifugation. The insert DNA was then injected into eggs of super-ovulating C57BL/6J mice at a concentration of 1-3 μg/ml. The microinjected embryos were transferred into pseudo-pregnant mice using standard procedure [11].

Tail clips from 3-week mice were used for DNA isolation. In founder animals the presence of transgene was established by Southern analysis of tail DNA using the 9.5 kb NheI-SacI human genomic fragment as a probe. In the offspring of the founders, the transgene was identified by PCR using the following primers VHL/ORI-1D: CTACCCAACGCTGCCGCCT and VHL/ORI-1R: AAAAAGAATGCTCTGACGCTTA. 25 cycles of PCR reaction were carried out at 94°C for 30 sec, 60°C for 30 sec and 72°C for 1 min.

Conditional Knockout Mice

Using *Cre/loxP* site-specific recombination technology, mice were generated with a targeted (*floxed*) allele in which exons 2 and 3 of the VHL gene were flanked by loxP sites [21]. The *floxed* VHL genomic region deleted by Cre recombinase activity encompassed 5.8 kb and started at nucleotide 5,696 of the mouse VHL sequence AF513984, located within intron 1. DNA isolated from tails of heterozygous F2 animals was used for bisulfite sequencing analysis of the VHL promoter region.

Animal care was provided in accordance with the procedures outlined in the "Guide for the Care and Use of Laboratory Animals" (NIH Publication No. 86-23, 1985).

Results

Long-range Bisulfite Sequencing of the Human VHL Locus

The genomic human VHL sequence has been deposited earlier, under GenBank accession numbers U97187 (sequence upstream of the VHL promoter, 4,761 bp) and AF010238 (VHL promoter, exons and introns, 14,543 bp). We first analyzed these sequences for the presence of repetitive elements using the CENSOR web server, which allows alignment of query sequence against a reference collection of human repeats [15]. Thirty-three full-length Alu repetitive elements were identified within the combined 19,304 bp of sequence, comprising almost half of it. Additional fragmented "fossil" repeats were also identified (Figure 1). Finally, the sequence was analyzed for the presence of potential matrix-(scaffold) attachment regions (MARs or SARs), using a program described elsewhere [30]. One statistically significant MAR was identified within VHL intron 1.

Human clear cell renal carcinoma cell line UMRC6 was selected for detailed methylation mapping of the 17 kb VHL locus. These cells express high levels of VHL mRNA, but encode a truncated VHL protein due to a 10 bp deletion in the coding region of exon 3 [19]. For bisulfite sequencing analysis the entire VHL locus was divided into 43 sub-regions, each amplified by nested PCR using primers specific for bisulfite-converted DNA (Table A1). The average length of the PCR fragments was 400 bp. Primers were placed between the Alu repeats avoiding CpG dinucleotides where possible. These fragments were sub-cloned into pCR 2.1 plasmid and 5 to 7 clones were sequenced for every sub-region.

Bisulfite sequencing revealed extensive methylation of the gene's body and the adjacent region upstream of the VHL promoter (Figure 1), implying that no regulatory elements associated with open chromatin conformation other than the VHL CpG island itself were present within the analyzed locus. Complete methylation was also evident in all "unique" sequences located between Alu repeats, including the coding regions of exons 2 and 3. The unmethylated area encompassed the 5' CpG island and the adjacent downstream region within intron 1, and was interrupted by a single partially methylated Alu repeat. The size of the protected area (including this Alu repeat) was 1.5 kb. Its upstream border coincided with the 3' end of the heavily methylated Alu repeat, so that the adjacent minimal VHL promoter [16] was free from methylation. The downstream border was located within a stretch of unique sequence in intron 1.

Comparative Methylation Analyses of the Mouse and Human VHL loci

The significance of some structural features revealed during the analysis of VHL locus can be ascertained by a comparison of orthologous loci between evolutionarily distant mammalian species. To compare the normal methylation patterns of the human and mouse VHL loci, we performed a detailed methylation mapping of the mouse VHL gene (Figure 2).

The full-length sequence of the mouse VHL gene was obtained as described in Materials and Methods.

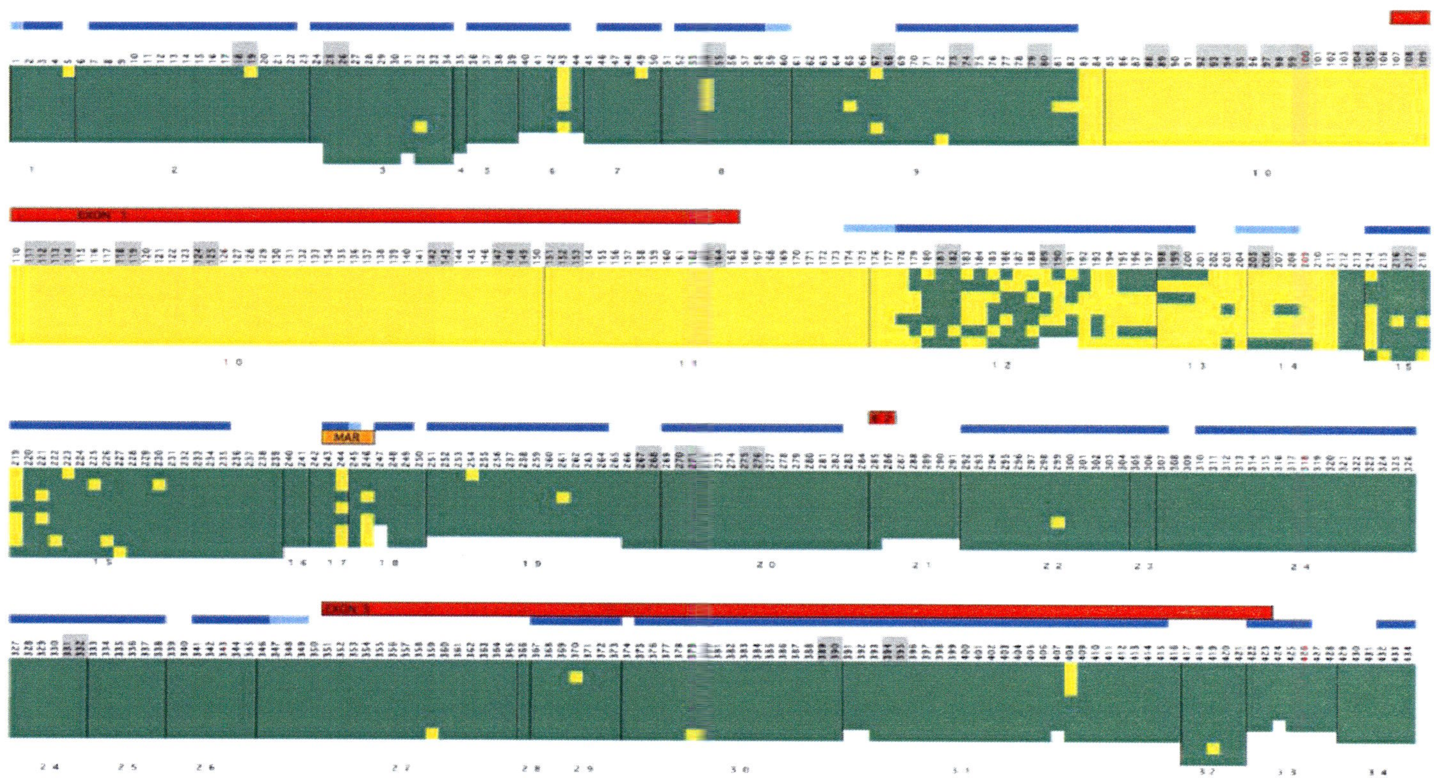

Figure 1. Methylation profile of the transcriptionally active human VHL locus in UMRC6 cells. Bisulfite sequencing of the PCR amplified products. For methylation analysis PCR fragments were subcloned and several plasmid clones were sequenced for each of the 43 sub-regions. Black vertical lines mark the borders between sub-regions. Each horizontal row in a sub-region delineates one plasmid clone. Yellow and green squares represent unmethylated and methylated CpG dinucleotides respectively. VHL exons are marked in red. Alu repeats are shown in dark blue, other repeats - in light blue. CpG dinucleotides located within BstUI cutting sites are marked by gray boxes. This profile covers 17 kb of genomic DNA and includes 434 CpG dinucleotides. First CpG dinucleotide of the profile corresponds to position 1345 of the DNA sequence [illegible]97187.

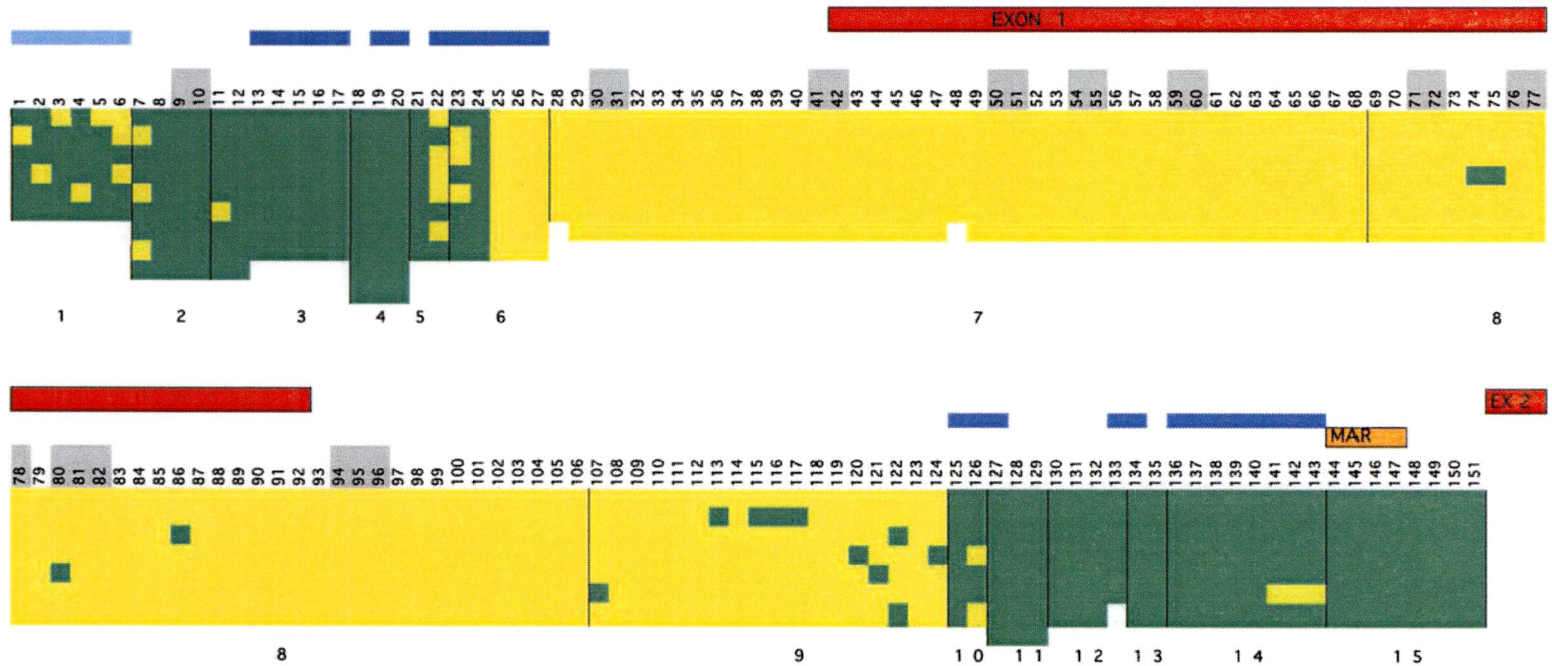

Figure 2. Methylation profile of the transcriptionally active mouse VHL locus in NIH-3T3 cells. Color designations are similar to Figure 1. CpG dinucleotides located within the Alu-like repetitive elements (B1) are marked by dark blue on the top. Other repetitive elements are marked in light blue. This profile covers 6.7 kb of genomic DNA and includes 151 CpG dinucleotide. First CpG dinucleotide of the profile corresponds to position 229 of the DNA sequence (submitted) mouse genomic VHL.

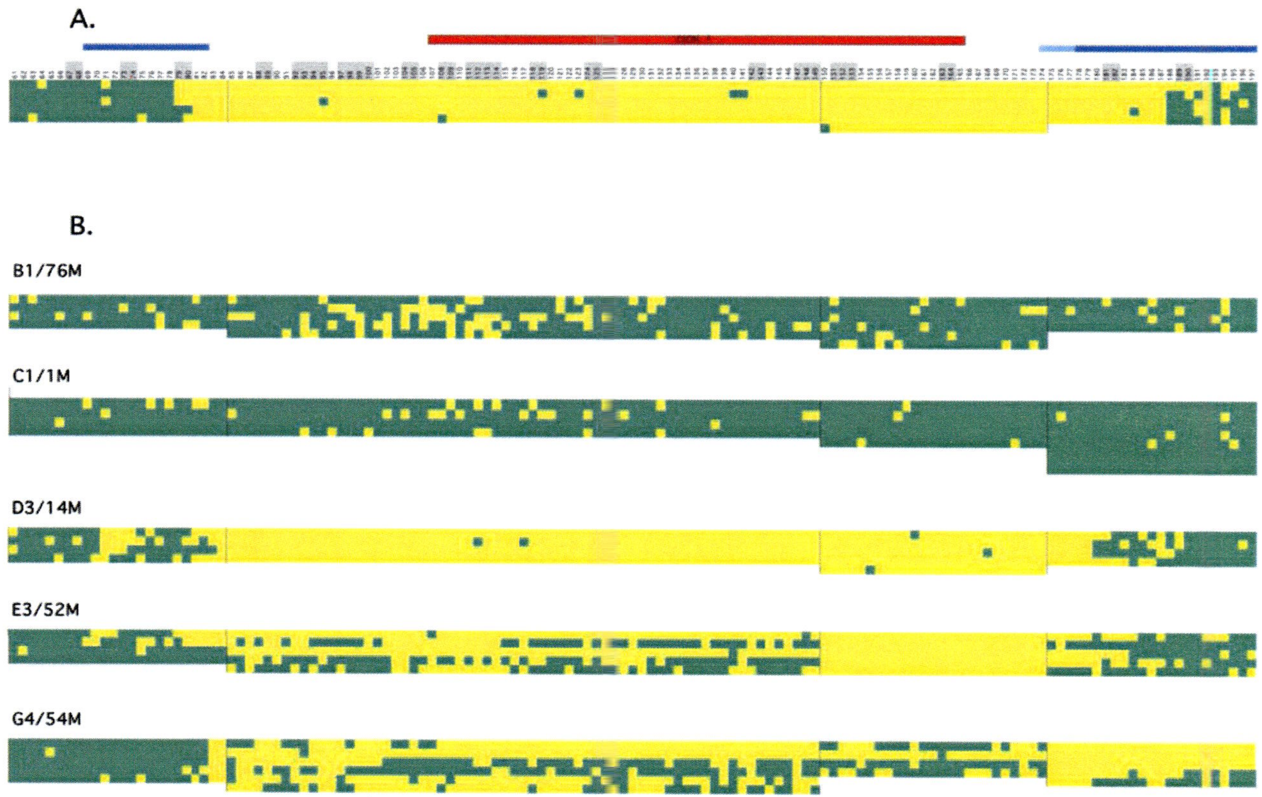

Figure 3. Methylation profiles of the human VHL transgenes. A. Methylation profile of the human VHL locus in the mouse A9 microcell hybrid MCH903.1. This cell line was kindly provided by Dr. E.J. Stanbridge. Color designations and CpG dinucleotide numbering are similar to Figure 1. B. Methylation profile of the human VHL transgenes in adult mouse. Color designations and the nucleotide numbering are the same as in Figure 1. Founder animal designations are shown on the left. The transgene includes 10 kb of human genomic DNA. This methylation profile covers 1479 bp of genomic DNA around CpG island and include 135 CpG dinucleotides.

Repetitive genomic sequences were mapped using BLAST homology search program and CENSOR repeat identification program [15]. In contrast to the human VHL locus, only 9 near full-length repeats belonging to the B1, B2, B3 family were identified within the 9 kb of available sequence. Additional "fossil" fragments of repetitive sequences were also found. Unique sequences were predominant in the mouse locus and comprised 72% of the analyzed 9 kb DNA. Similar to human VHL locus, one MAR was predicted within intron 1 of the mouse gene. Despite the presence of Alu repeats in human VHL gene, the size of intron 1 was similar in the mouse and human VHL loci.

Next, we asseembled a detailed CpG methylation map of the 6.7 kb genomic region surrounding the mouse VHL CpG island, including intron 1, exon 1 and 2.5 kb upstream of exon 1. Bisulfite sequencing was performed using DNA from NIH 3T3 cells, which express high levels of VHL mRNA. Bisulfite-treated DNA was amplified by PCR in 15 separate fragments. Each PCR product was sub-cloned into pCR2.1 and 5 to 7 clones were sequenced for every sub-region. Notwithstanding the absence of repetitive elements and great divergence of the sequence outside VHL coding region, the methylation patterns of the mouse VHL gene and human locus were very similar. A CpG island (1.0 kb) and a short relatively GC-poor region immediately downstream (0.5 kb) were essentially free from methylation (Figure 2). The total size of the protected area was identical to that of the human VHL locus and comprised about 1.5 kb. Other sequences upstream (2.3 kb) and downstream (3.0 kb) of the protected area were completely methylated.

Global human-mouse VHL sequence alignment was performed using "Vista" programs [22]. Apart from similarities between human Alu and mouse B1, only one short region of homology (89 bp, 80%) was found within intron 1 (starting at the nucleotide 2,951 of the human AF010238 sequence and nucleotide 5,287 of the mouse sequence (submitted)). The same region of homology was identified using the "PipMaker" homology search program [28].

Since a putative matrix attachment region, similar to human MAR, was found in intron 1 of the mouse gene; we studied its significance for the control of local methylation patterns. We analyzed heterozygous conditional knockout mice in which this site was deleted in one of the two alleles (see Materials and Methods). No perturbations of methylation patterns were found in the affected alleles (data not shown).

Pre-existing Methylation of the Human VHL Locus is Properly Maintained in Mouse Somatic cells

The numerous structural similarities found between the mouse and human VHL loci prompted us to investigate whether the mouse methylation machinery can maintain pre-existing methylation patterns on a human DNA template. We analyzed methylation of the human VHL locus in a mouse-human microcell hybrid MCH903.1 containing human chromosome 3 as a sole human genetic component [13]. The bisulfite-treated genomic DNA from this cell line was amplified in four separate nested PCR reactions covering a continuous region, which include the VHL CpG island and flanking genomic sequences. This region started at nucleotide 50 of the sequence AF010238 and included CpG dunucleotides from 61 to 197 (according to the numeration on figure 1). In MCH903.1 cells, the methylation pattern around the CpG island was almost identical to that found in UMRC6 human renal carcinoma cells (Figure 3A).

Methylation of the Human VHL Locus in Transgenic Mice

Since *de novo* methylation of transgenes does not occur in human or rodent somatic cells grown in culture, we studied whether the VHL methylation patterns can be faithfully reproduced during development in other species. For this purpose, we created mice carrying the human VHL transgene.

This transgene included a human genomic NheI/SacI fragment (9,963 bp) containing the 5' part of the VHL gene. The fragment was subcloned into pBluescript II SK from phage P-191 [19]. It included a 2.5 kb region upstream of the VHL promoter, the first and second exons, intron 1 and a part of intron 2. The insert DNA was excised and injected into oocytes of C57BL/6J mice. DNA was isolated from tails of developed animals and analyzed for presence of the human VHL transgene by Southern hybridization (Figure 4). At least five founder mice carrying a non-rearranged VHL transgene were identified. These mice were bred with wt C57 mating partners, thus subjecting the transgene to a full array of developmental chromatin- and DNA-modification factors in F1 progeny. DNA from tails was isolated and analyzed by PCR using human VHL-specific DNA primers to identify animals carrying the transgene. One animal from each of the five transgenic lineages was analyzed by bisulfite sequencing of the human VHL transgene. The bisulfite-treated DNA from tails was amplified in four separate nested PCR reactions covering the continuous region, which included the CpG island and flanking genomic sequences (Fig. 3B). Total length of the bisulfite-amplified part of transgene was 1660 bp. The amplified region was identical to that of the MCH903.1 mouse-human microcell hybrid. The PCR products were cloned and sequenced (four to eight clones per region).

We found heavy methylation of the sequences upstream from the VHL promoter in every analyzed animal. In offspring animals originated from founder mice B1 and C1, a dense methylation of the entire transgene was evident (Figure 3B). However, in the progeny of founder mice D3, E3 and G4 the transgenic human VHL CpG island was protected from methylation. In these three lineages the methylation pattern of the transgene was similar to that found in human cells expressing VHL mRNA: unmethylated CpG island surrounded by densely methylated adjacent regions.

We analyzed four different animals for each of the five lineages (total of 20) by combined bisulfite-restriction analysis (COBRA), using the same PCR primers as we did for bisulfite sequencing. All four PCR fragments contain BstUI and/or TaqI restriction sites that include CpG dinucleotides in their recognition sequences. These CpGs would remain preserved in bisulfite-treated DNA only if methylated in the original DNA. Therefore, BstUI or TaqI cutting of the fragments would indicate DNA methylation. According to COBRA analysis, in all the B1 and C1 progeny animals methylation was present throughout the entire VHL transgene. In D3, E3 and G4 lineages, the island but not the flanking genomic sequences was protected from methylation (data not shown). These results are in agreement with the data we obtained earlier using bisulfite sequencing of five lineage-representative animals (shown on figure 3B).

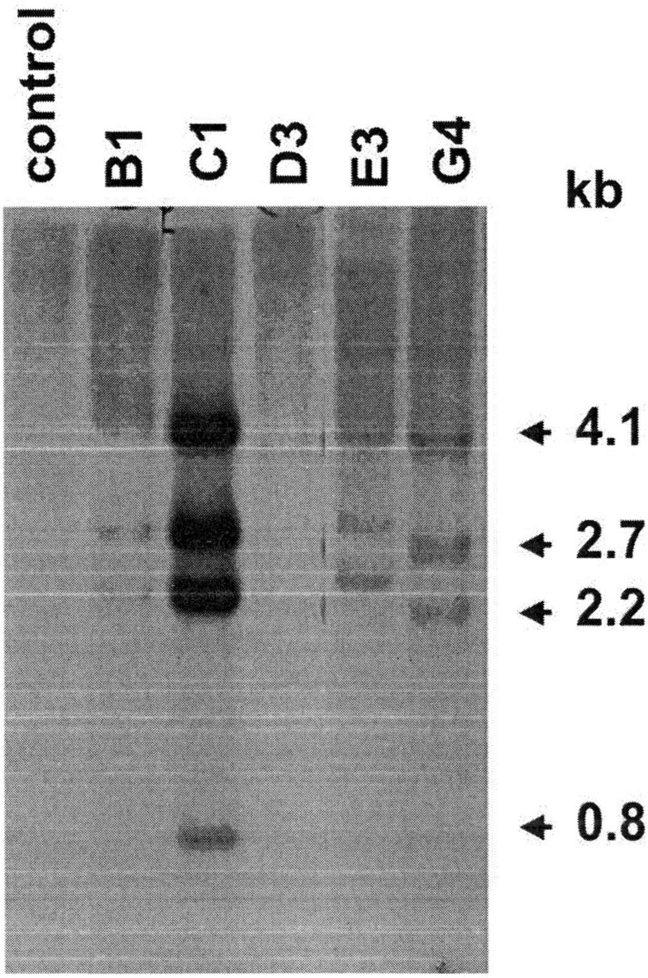

Figure 4. Southern analysis of the VHL transgenic founder animals. Restriction digest with BamHI. Similarity in restriction patterns confirms integrity of human VHL transgenes.

Discussion

The pivotal role of DNA methylation in the regulation of gene expression in mammals has been discussed for some time [3]. Methylation marks create an additional level of hierarchy in complex vertebrate genomes. According to some suggestions, this marking prioritizes regulatory sequences in regard to their availability for protein factors regulating gene expression [3], creating differentially accessible chromatin conformations. Indeed, experimental methylation of promoters inhibits transcription. Another function of DNA methylation, related to the first, is its association with suppression of the spurious proliferation of "selfish" DNA repetitive elements [32]. The question, however, remains if this suppressive methylation is limited exclusively to the repetitive elements themselves or able to spread to adjacent "unique" genomic sequences. In the latter case, the functioning genomic elements would have to be protected from methylation by locus-specific mechanisms. Such mechanisms have already been proposed for the *aprt* promoter [31] and several loci controlled by a highly versatile CTCF protein factor [24] mutated in different tumor types [5]. These proteins bind their DNA recognition sequences and function as chromatin insulators, separating condensed and open chromatin. Whether similar methylation protection mechanisms are present in other loci, specifically those implicated in carcinogenesis, remains to be determined. To support this hypothesis, long-range methylation maps of the critical loci must be generated. These maps should comprise relatively long stretches of genomic DNA including CpG islands, adjacent interspersed repetitive elements and "unique" sequences.

In this work we assembled a high-resolution methylation map of the entire human VHL locus extending 17 kb, to our knowledge, the largest continuous methylation profile created so far. The results indicate that bisulfite sequencing map construction is feasible for the extended genomic loci, regardless of the presence of large numbers of repetitive elements and decreased complexity of DNA sequence after bisulfite treatment. We also assembled a similar map of 6.7 kb around the mouse VHL CpG island. Despite considerable divergence of genomic sequences in the two VHL loci, we found similar methylation patterns in the human and mouse. In both genes 1.5 kb promoter-containing regions were protected from methylation, including the CpG islands and adjacent 0.5 kb portion of intron 1. Other sequences, both repetitive and unique were heavily methylated in these loci.

As for all other mammalian loci, the spatial interrelationships of DNA methylation and histone modification profiles of the VHL gene remain to be elucidated. Recent data indicate that at least in yeast cells, the pivotal methylation of H3 histone at Lys^4 is not specifically limited to the 5' regions of the genes and extends to much larger euchromatic compartments [23]. Similar chromatin organization in higher eukaryotes may indicate that DNA methylation and histone modification patterns are not completely identical and point to an additional level of hierarchy in the chromatin.

The upstream border of the VHL minimal promoter [16] coincided with the beginning of the methylation-protected region, indicating that some of the promoter-binding proteins may perform protecting function. Mechanisms of the downstream border formation are less clear. It could be created as the result of binding by other sequence specific protein factors. Alternatively, the active unidirectional VHL promoter may somehow maintain an open chromatin conformation for some distance (limited to 1.5 kb) in the direction of transcription. In any case, the factors affecting the length of the unmethylated sequence remain unknown. Deletion analysis of the protected region performed in a *de novo* methylation environment may resolve this question.

According to one model, only small regions including the promoters of expressed genes are located within the specific chromatin milieu permissive for interaction with gene-specific transcription factors [9]. The remaining parts of the genes and, probably, most of the intergenic regions are immersed in more compacted restrictive chromatin, associated with predominantly methylated DNA. For this reason most of the genomic DNA is insulated from interaction with DNA-binding promoter-specific protein factors. This model would explain an apparent discrepancy between the large number of consensus DNA binding sequences found in the available genomic sequence databases and the limited number of active gene-regulatory elements. For now *in silico* promoter prediction remains unreliable [33]. Better predictive values can be achieved by superimposition of methylation profiles on the map of consensus binding sites.

To study whether the VHL-specific methylation-protection mechanisms are functional across different species, we created several independent lineages of transgenic mice carrying a 10 kb human genomic region which includes the CpG island. In 3 of 5 lineages, the methylation patterns of the human transgene closely resembled the normal pattern found in human cells expressing VHL. In at least one lineage (C1), hypermethylation of the human transgene may be explained by its higher copy number (Figure 4). This data indicate that all essential human DNA target sequences involved in VHL-protection are present within the transgene and interchangeable between mouse and human. The ability of developing mice to form normal methylation patterns on human DNA templates makes possible systematic dissection of the individual elements relevant for formation and maintenance of methylation.

As we mentioned, the human VHL CpG island was protected from methylation in three successive generations of transgenic mice in the absence of selective pressure, despite heavy methylation of the adjacent sequences. Methylation of delivered transgenes represents a considerable problem in the various gene therapeutic applications [1,12,25]. We propose that similar protective sequences could be included in the vectors used for gene therapy.

Acknowledgments

This publication has been funded in part with Federal funds from the National Cancer Institute, National Institutes of Health, under contract No. NO1-CO-12400 and NO1-CO-56000.

The content of this publication does not necessarily reflect the views or policies of the Department of Health and Human Services, nor does mention of trade names, commercial products, or organizations imply endorsement by the U.S. Government.

References

[1] Bestor, T.H. 2000. Gene silencing as a threat to the success of gene therapy. *J. Clin. Inv.*, *105*: 409-411.

[2] Bird, A. 2002. DNA methylation patterns and epigenetic memory. Genes Dev., *16*: 6-21.

[3] *Bird, A.P.* 1986. CpG-rich islands and the function of DNA methylation. *Nature*, *321*: 209-213.

[4] Costello, J.F., Plass, C. 2001. Methylation matters. *J. Med. Genet.*, *38*: 285-303.

[5] Filippova, G.N., Qi, C.F., Ulmer, J.E., Moore, J.M., Ward, M.D., Hu, Y.J., Loukinov, D.I., Pugacheva, E.M., Klenova, E.M., Grundy, P.E., Feinberg, A.P., Cleton-Jansen, A.M., Moerland, E.W., Cornelisse, C.J., Suzuki, H., Komiya, A., Lindblom, A.,

Dorion-Bonnet, F., Neiman, P.E., Morse, H.C. 3rd, Collins, S.J., Lobanenkov, V.V. 2002. Tumor-associated zinc finger mutations in the CTCF transcription factor selectively alter tts DNA-binding specificity. *Cancer Res.*, *62*: 48-51.

[6] Graff, J.R., Herman, J.G., Myohanen, S., Baylin, S.B., Vertino, P.M. 1997. Mapping patterns of CpG island methylation in normal and neoplastic cells implicates both upstream and downstream regions in de novo methylation. *J. Biol. Chem.*, *272*: 22322-22329.

[7] Grossman, H.B., Wedemeyer, G., Ren, L.Q. 1985. Human renal carcinoma: characterization of five new lines. *J. Surg. Oncol.*, *28*: 237-244, 1985.

[8] Grunau, C., Clark, S.J., Rosenthal, A. 2001. Bisulfite genomic sequencing: systematic investigation of critical experimental parameters. *Nucl. Acids Res. 29*: e65.

[9] Grunau, C., Hindermann, W., Rosenthal, A. 2000. Large-scale methylation analysis of human genomic DNA reveals tussue-specific differences between the methylation profiles of genes and pseudogenes. *Human Mol. Gen.*, *9*: 2651-2663

[10] Herman J.G., Latif, F., Weng, Y., Lerman, M.I., Zbar, B., Liu, S., Samid, D., Duan, D.S., Gnarra, J.R., Linehan W.M., Baylin, S.B. 1994. Silencing of the VHL tumor suppressor gene by DNA methylation in renal carcinoma. *Proc. Natl. Acad. Sci. USA*, *91*: 9700-9704.

[11] Hogan, B., Beddington, R., Costantini, F., Lacy, E.1994. *Manipulating the mouse embryo. A laboratory manual. 2nd Ed,* Cold Spring Harbor Press

[12] Hong, K., Sherley, J., Lauffenburger, D.A. 2001. Methylation of episomal plasmids as a barrier to transient gene expression via a synthetic delivery vector. *Biomol. Eng.*, *18*: 185-192.

[13] Imreh, S., Kholodnyuk, I., Allikmets, R., Stanbridge, E.J., Zabarovsky, E.R., Klein, G. 1994. Nonrandom loss of human chromosome 3 fragments from mouse-human microcell hybrids following progressive growth in SCID mice. Genes Chrom. *Cancer*, *11*: 237-245

[14] Jainchill, J.L., Aaronson, S.A., Todaro, G.J. 1969. Murine sarcoma and leucemia viruses: assay using clonal lines of contact-inhibited mouse cells. *J. Virol.*, *4*: 549-553.

[15] Jurka, J., Klonowski, P., Dagman, V., Pelton, P. 1996. CENSOR – a program for identification and elimination of repetitive elements from DNA sequences. *Computers and Chemistry*, *20*: 119-122.

[16] Kuzmin, I., Duh, F.-M., Latif, F., Geil, L., Zbar, B., Lerman, M.I. 1995. Identification of the promoter of the human von Hippel-Lindau disease tumor suppressor gene. *Oncogene. 10*: 2185-2194.

[17] Kuzmin, I., Geil, L. 2001. DNA methylation and chromatin modifications in cancer and development. *Int. Arch. Biosci., 1*:1047-1056.

[18] Kuzmin, I., Geil, L., Ge, H., Bengtsson, U., Duh, F.-M., Stanbridge, E.J., Lerman, M.I. 1999. Analysis of aberrant methylation of the VHL gene by transgenes, monochromosome transfer and cell fusion. *Oncogene. 18*: 5672-5679.

[19] Latif, F., Tory, K., Gnarra, J., Yao, M., Duh, F.-M., Orcutt, M.L., Stackhouse, T., Kuzmin, I., Modi, W., Geil, L., Schmidt, L., Zhou, F., Li, H., Wei, M.H., Chen, F., Glenn, G., Choyke, P., Walther, M.M., Weng, Y., Duan, D.-S.R., Dean, M., Glavac, D., Richards, F.M., Crossey, P.A., Ferguson-Smith, M.A., Le Paslier, D., Chumakov, I.,

Cohen, D., Chinault, A.C., Maher, E.R., Linehan, W.M., Zbar, B, Lerman, M.I. 1993. Identification of the von Hippel-Lindau disease tumor suppressor gene. *Science*, *260*: 1317-1320.

[20] Lorincz, M.C., Schubeler, D., Goeke, S.C., Walters, M., Groudine, M., Martin, D.I.K. 2000. Dynamic analysis of proviral induction and de novo methylation: Implications for a histone deacetylase-independent, methylation density-dependent mechanism of transcriptional repression. *Mol. Cell. Biol. 20*: 842-850.

[21] Ma W, Tessarollo L, Hong SB, Baba M, Southon E, Back TC, Spence S, Lobe CG, Sharma N, Maher GW, Pack S, Vortmeyer AO, Guo C, Zbar B, Schmidt LS. 2003. Hepatic vascular tumors, angiectasis in multiple organs, and impaired spermatogenesis in mice with conditional inactivation of the VHL gene. *Cancer Res.* 2003 *63*:5320-8.

[22] Mayor, C., Brudno, M., Schwartz, J.R., Poliakov, A., Frazer, K.A., Pachter, L.S., Dubchak, I. 2000. Visualizing global DNA sequence alignments of length. *Bioinformatics,* 16: 1046.

[23] Noma, K., Allis, C.D., Grewal, S.I.S. 2001. Transitions in distinct histone H3 methylation patterns at the heterochromatin domain boundaries. *Science*, *293*: 1150-1155.

[24] Ohlsson, R., Renkawitz, R., Lobanenkov, V. 2001. CTCF is a uniquely versatile transcription regulator linked to epigenetics and disease. *Trends Genet.*, *17*: 520-527.

[25] Pannell, D., Ellis, J. 2001. Silencing of gene expression: implication for design of retrovirus vectors. *Rev. Med. Virol.*, *11*: 205-217.

[26] Pierce, J.C., Sternberg, N., Sauer, B. 1992. A mouse genomic library in the bacteriophage P1 cloning system: organization and characterization. Mammalian Genome, *3*: 550-558.

[27] Rougier, N., Bourc'his, D., Gomes, D.M., Nivelau, A., Plachot, M., Paldi, A., Viegas-Pequignot, E. 1998. Chromosome methylation patterns during mammalian preimplantation development. *Genes Dev.*, *12*: 2108-13.

[28] Schwartz, S., Zhang, Z., Frazer, K.A., Smit, A., Riemer, C., Bouck, J., Gibbs, R., Hardison, R., Miller, W. 2000. PipMaker – A Web Server for aligning Two Genomic DNA Sequences. *Genome Res.*, *10*: 577-586

[29] Siegfried, Z., Eden, S., Mendelsohn, M., Feng, X., Tsuberi, B.-Z., Cedar, H. DNA methylation represses transcription in vivo. 1999. *Nature Genetics*, *22*: 203-206.

[30] Singh, G.B., Kramer, J.A., Krawetz, S.A. 1997. Mathematical model to predict regions of chromatin attachment to the nuclear matrix. Nucl. *Acids Res.*, *25*: 1419-1425.

[31] Turker, M.S., Mummaneni, P., Cooper, G.E. 1995. The mouse *aprt* gene as a model for studying epigenetic gene inactivation. *Adv. Exp. Med. Biol.*, *370*: 647-652.

[32] Yoder, J.A., Walsh, C.P., Bestor, T.H. 1997. Cytosine methylation and the ecology of intragenomic parasites. *Trends Genet.*, *13*: 335-340.

[33] Zhang, M. Q. 1998. Identification of human gene core promoters in silico. *Genome Res.*, *8*: 319-326.

Appendix

Table A1. Primers for bisulfite-treated human VHL locus

PCR FRAGMENT	PCR RXN	FORWARD PRIMER	SEQUENCE 5'-->3'	REVERSE PRIMER	SEQUENCE 5'-->3'
1	1	BIS-14D	ATTTGTAGATAC/TGAATGTGGAG	BIS-14CR	ACTAATAACTTATTATATA
	2	BIS-15D	TATTTGGAATAGTTATTTATC/TGTTGG	BIS-14BR	TTTTACCTTATTTATATATCAACC
	3	BIS-14AD	GAGATTGTAAAATGATATAATTGTTTTGGGGGA	BIS-14AR	ACAAATACTAATCATATCACATATA
2	1	BIS-73D	TAATTATGTGATATGATTAGTATTTG	BIS-73R	AATACACAACCAAAATTTTTCTTACAA
	2	BIS-74D	TATTATTGTATTTGTATATTTTAATTGGT	BIS-74R	TACACTTTTCATTAACCAAAAAAACAT
3	1	BIS-71D	ATGTTTTTTTGGTTAATGAAAACTGTA	BIS-71R	ATATAACAATTTAATATTACAAAAATACC
	2	BIS-72D	TTGTAAGAAAAATTTTGGTTGTCTATT	BIS-72R	TACATCCCTAAAATAAAACCAA
4	1	BIS-69D	AATAATTGGTTTTATTTTAGGGATGTA	BIS-69R	AATTCACAAAATTCATACAAACATCC
	2	BIS-70D	GGTATTTTTGTAATATTAAATTGTTATAT	BIS-70R	CCTCAATTTTACTCCCATACTAAC
5	1	BIS-67D	GGGGATGTTTGTATGAATTTTGTG	BIS-67R	AACCAACAATAATACTAAACCCCC
	2	BIS-68D	AGAAGTTTTATTTTGGGGTTTATTTTA	BIS-68R	TACATTAACAACATAAAAACACTATA
6	1	BIS-65D	GTTTAATTTTTGAGGTTTATTTATTATTG	BIS-65R	CTATATCATTATCTCCATTACACAAC
	2	BIS-66D	GATGAGTTTTTTTGATTTAGTATTTTG	BIS-66R	ATAAAAATATAAAAACCTAACACTATAC
7	1	BIS-63D	GTATAGTGTTAGGTTTTTATATTTTTA	BIS-63R	ATTAAATATTCAATAATATAAAATAAACTC
	2	BIS-64D	ATGGAGATAATGATATAGTTTTAATTTA	BIS-64R	AAACTCAAATAAAAAAACAACTCAAAAA

Table A1. Primers for bisulfite-treated human VHL locus (Continued)

PCR FRAGMENT	PCR RXN	FORWARD PRIMER	SEQUENCE 5'-->3'	REVERSE PRIMER	SEQUENCE 5'-->3'
8	1	BIS-61DA	GAGTTTATTTATATTATTGAATATTTAAT	BIS-61R	ATATTTACTTCAAAACAATTAAAACTAC
	2	BIS-61D	TATAGAATTGTGTATTATTTTTTTGATT	BIS-62R	AACATAAACACTCATTAACTTTATTCA
	3	BIS-62D	GATTTTTAGTTTTTATAGTATTTTGTT	BIS-62RA	AACCCTACTATATACCACCTATAAT
9 (frag. 0)	1	BIS-5D	AATTTAGTTAGTTGATTTTTTGTATT	BIS-13R	CTCG/AATAAAAAATAAAACG/ACG/ACTC
	2	BIS-6D	ATAAGC/TGTGATGATTGGGTG	BIS-5R	ATAACCG/AACCG/AAAAACTATAC
	3	BIS-13D	GTGTTTTC/TGTGTGAGATG	BIS-4R	CAAAAAAATCCTCCAACACC
10 (frag. 1)	1	BIS-21D	AGTGGAAATATAGTAAGAGTTGGT	BIS-79R	TCG/AAAATTAAACCATACG/AAACAACAC
	2	BIS-4D	GAGTTGGTTTAGTTTC/TGTTT	BIS-80R	ATTACAAAAAATAACCTAAAAAAACTC
11 (frag. 2)	1	BIS-77D	GC/TGTC/TGAGGAGGAGATGGAGGT	BIS-77R	AACAACG/ATCACCCTAAATATATCC
	2	BIS-78D	GAGTTTTTTTAGGTTATTTTTTGTAAT	BIS-78R	ACCTCAAAAAACCTCAATTCCCC
12 (frag. 3)	1	BIS-11D	GATTAGTAGGGAC/TGATAGTA	BIS-75R	G/AACCTTATAAAAAATACTCTAAACC
	2	BIS-12D	GTTTC/TGGGGTTTATTTTGTAGA	BIS-76R	CATAAACCCAACACTTTAAAAAAACC
13	1	BIS-16D	GAATAGTTTGTATTTGAAGGTG	BIS-16R	TCACAACCCATATTCCTAAAA
	2	BIS-17D	TAGTTTTTATGATTTTAGTGGG	BIS-17R	AATCACCTTTTTAAACATCCTCTC
14	1	BIS-21D	GCGTTGAGATTTGGAGTTTTGG	BIS-21R	TTTCCCCTCCAAACCACTTAAC
	2	BIS-22D	GAGAGGATGTTTAAAAGGTGATT	BIS-22R	AAATAAACTTTTCAATATATAAATCTAAC
	2	BIS-22AD	TTTAGGAATATGGGGGTATTGAAAT	BIS-22RA	ACAATATCATATCCCTACTTAAAAAC
15	1	BIS-37D	AGTTTTTAAGTAGGGATATGATATTG	BIS-37R	TCCCTAAAAATACTATATTCTTCATC
	2	BIS-38D	GAAAAGTTTATTTAGGTTGTTAAGTG	BIS-38R	AATATATTAAATACACCACCCTCAAA

Table A1. Primers for bisulfite-treated human VHL locus (Continued)

PCR FRAGMENT	PCR RXN	FORWARD PRIMER	SEQUENCE 5'-->3'	REVERSE PRIMER	SEQUENCE 5'-->3'
16	1	BIS-39D	TTTGAGGGTGGTGTATTTAATA[illegible]TT	BIS-39R	ATACCTATAATCCCAACTACTTTAAAA
	2	BIS-40D	GATGAAGAATATAGTATTTTTA[illegible]GGA	BIS-40R	AAACTCTAAAAAAACTACTTAAACCC
	3	BIS-40DA	GTAGGTTTTTATTTTATATAGGA[illegible]ATTG	BIS-40RA	AACCCAAATTAAAAATACAATAAACC
17	1	BIS-41D	GGGTTTAAGTAGTTTTTTTAGAG[illegible]TT	BIS-41R	TTTCAAAATACATACATCACAAACAAA
	2	BIS-42D	AGTTTTTTAAAGTAGTTGGGAT[illegible]TAG	BIS-42R	CTCCAACCCTAAAATTCTATTTAAAA
	3	BIS-42AD	GGTATGTTTTATTATGTTTGGTT[illegible]ATT	BIS-42AR	TCAAAATAAATATAACACTTATTCCTC
18	1	BIS-43DA	GAGGAATAAGTGTTATATTTATT[illegible]TGAG	BIS-43R	CCCTTCTAATAACATTATCTCAATC
	2	BIS-43D	TTTTAAATAGAATTTTAGGGTTG[illegible]AG	BIS-44R	CTCACAACAAATCCATTACAAATAA
	3	BIS-44D	TTTGTTTGTGATGTATGTATTTT[illegible]AAA	BIS-44RA	CATACTCAAAATATAAAACAATAAACAA
19	1	BIS-45D	TGTATAATATTTAAGTAGAGTG[illegible]GAG	BIS-45R	TCAATTTAATTAACCAAAAACCCATC
	2	BIS-46D	GGTTAGGGGAAATGGAGAAAA[illegible]G	BIS-46R	TATATCCCTACATCTCTAAAAAACC
20	1	BIS-47DA	AATTTTTGTTTGTTTC/TGATAGG[illegible]TATT	BIS-47R	CAACAACAAATAAACCAAAATATCTC
	2	BIS-48	TGGGTTTTTGGTTAATTAAATTC[illegible]	BIS-48R	ATACTCTACACATTATTCTCCTAAAA
21	1	BIS-49R	TTTTAGGAGAATAATGTGTAGA[illegible]TAT	BIS-49R	TTACCCACAAAAACATCTAATTTATTAA
	2	BIS-5OR	GAGATATTTTGGTTTATTTGTTC[illegible]TG	BIS-50R	CCCTAATAAAATCATTTAAACTTCAAA
22	1	BIS-51D	TTTGAAGTTTAAATGATTTTATT[illegible]GGG	BIS-51R	CTTCCTTAAAACAAAATCTATACAC
	2	BIS-52D	TTAATAAATTGATGTTTTTGTGC[illegible]TAA	BIS-52R	CATAATAATTTAAATCCTATAAAACCC
23	1	BIS-53D	GGGTTTTATAGGATTTAAATTA[illegible]TATG	BIS-53R	AAATAATAATCACAAAACACATAACCA
	2	BIS-54D	GTGTATAGATTTTGTTTTAAGGA[illegible]G	BIS-54R	AACAATATATACTTAAAAAATCTAAACC

Table A1. Primers for bisulfite-treated human VHL locus (Continued)

PCR FRAGMENT	PCR RXN	FORWARD PRIMER	SEQUENCE 5'-->3'	REVERSE PRIMER	SEQUENCE 5'-->3'
24	1	BIS-55D	GGTTTAGATTTTTTAAGTATATATTGTT	BIS-55R	ACTACCACCAACAAAATAACTACC
	2	BIS-56D	TGGTTATGTGTTTTGTGATTATTATTT	BIS-56R	CTAAACCCAACCTAAACTACAAAA
25	1	BIS-57D	TTTTGTAGTTTAGGTTGGGTTTAG	BIS-57R	ACCTACTAACTTCTACTTTTATATTT
	2	BIS-58D	GGTAGTTATTTTGTTGGTGGTAGT	BIS-58R	AAAAACCCTAAACTACCCTACTAAA
26	1	BIS-59D	TTTAGTAGGGTAGTTTAGGGTTTTT	BIS-59R	CAAAACTCATCAATACCATCAAAAAC
	2	BIS-60D	AAGTATAAAAGTAGAAGTTAGTAGGT	BIS-60R	TCAACAAAAATCTTCAATCTCCCAT
27	1	BIS-35D	GTTTGTATGTAAGGAGGTTTATAT	BIS-35R	ACAAAAACTTATTTATCCATTAATAC
	2	BIS-36D	GTAATTTAGTGGGAATTGTAGTAT	BIS-36R	CTATAATCAAACATACAAATTTTCTC
28	1	BIS-33DA	ATATGTGATATTTTTGATTGATTTGGG	BIS-33RA	CTAAATCTAACCTATAAAAATTTATACA
	2	BIS-33D	TTTGGGATGGAGTTTTATTTTTGT	BIS-33R	CTCCATAAATCCATTAATAAAACC
	3	BIS-34D	TATTTAGGTTGGAGTGTAGTGG	BIS-34R	TAAAAATCCACCTAATCCTACC
29	1	BIS-31D	TGTTTGTGGGTATTTTTTGGTATG	BIS-31R	TATTAAATATTAAAAAAAAAAACTAAAATTC
	2	BIS-32D	TTAGTTTTTGGGTATAATTTTTGTG	BIS-32R	ATATCTCATTCACTTCACTCTAAT
30	1	BIS-29D	ATTAGAGTGAAGTGAATGAGATAT	BIS-29R	CCTACAATTTTAACAAAAAAAACAT
	2	BIS-30D	GAATTTTAGTTTTTTTTTTAATATTTAATA	BIS-30R	ACTATCATTTTAACACAAAAACAAC
31	1	BIS-27D	GTTGTTTTTGTGTTAAAATGATAGT	BIS-27R	ACAAATTTTAATTTTCCTTAAACAAC
	2	BIS-28D	ATGTTTTTTTTGTTAAAATTGTAGG	BIS-28R	CACCAAAAACACAATACTCAAAC

Table A1. Primers for bisulfite-treated human VHL locus (Continued)

PCR FRAGMENT	PCR RXN	FORWARD PRIMER	SEQUENCE 5'-->3'	REVERSE PRIMER	SEQUENCE 5'-->3'
32	1	BIS-25D	GTTTGAGTATTGTGTTTTTGGTG	BIS-25R	ACCTTATTCTCAACATAATTACTAA
	2	BIS-26D	GTTGTTTAAGGAAAATTAAAATTTTGT	BIS-26R	TAAAATACTACCACTTACCCTTT
33	1	BIS-23D	AAAGGGTAAGTGGTTAGTATTT[illegible]A	BIS-23R	CATTAAACTCTTAAATTTTAAAATAC
	2	BIS-24D	TTAGTAATTATGTTGAGAATAAC[illegible]GT	BIS-24R	TCAAAAAACCAATAACTAACAAAAC
34	1	BIS-18D	GTATGGGTTAGAAATTAGTGTA	BIS-18R	AATTCAACTATCCTAACACTCT
	2	BIS-19D	TC/TGGATATTTTGATTTTGGGTTT	BIS-19R	TTATAATACACCATTCCTTCC
	3	BIS-20D	GTTTTAC/TGTTTTGTTAGTTATTT[illegible]G	BIS-20R	CCCAAAATATAAAATAAAACCTTCT

Table A2. Primers for bisulfite-treated mouse VHL locus

PCR FRAGMENT	PCR RXN	FORWARD PRIMER	SEQUENCE 5'-->3'	REVERSE PRIMER	SEQUENCE 5'-->3'
1	1	BIM-31D	GGAATATTTTTTTTTTGATAGAGTA	BIM-31R	CCAATCTAAAAAATTCAATACCC
	2	BIM-32D	GTAATGTTAATTGTTGAGTTTTTTTT	BIM-32R	CTAACCTCCACAAACATATACAA
2	1	BIM-29D	TGTATTTTTTAATTGTTTTAAAAGTATG	BIM-29R	AATTCTCTTAAATTTTTAATTCCCC
	2	BIM-30D	TATAAATTTTTTATTGTTTAGGGTGAT	BIM-30R	AAAATCCAAAACTTTATATATAACAC
3	1	BIM-27D	TTTGGAGAGTTGATTTAGTGGTT	BIM-27R	CAATACCATTTTTCCATAAACATTT
	2	BIM-28D	TAGTTTTTATAATAGGTAGTTTATAG	BIM-28R	CTAACAAAAATTTCATCAAATAAACT
4	1	BIM-25D	GGTTAGTTTGGGTTATATATATTTA	BIM-25R	AACTTACCATATTTCCCCTTCTA
	2	BIM-26D	AGGTAATGGGAATTATGTAATAAG	BIM-26R	TATTAATATAAAAACTAAATACCACC
5	1	BIM-23D	TTTGAAAAATTAAAAAGTAAAGGTAG	BIM-23D	AACTCAAAAATCTCCCTACCTCAA
	2	BIM-24D	TAGAAGGGGAAATATGGTAAGTT	BIM-24D	CTCCAAAAAACAAAAACCAAAAACAA
6	1	BIM-1D	TTGTTTTTGGTTTTTGTTTTTTGGAG	BIM-1R	AAATTATAAATTCCACCTCCTACAAT
	2	BIM-2D	TTGAGGTAGGGAGATTTTTGAGTT	BIM-2R	CAATTAAATTTCTATTATTTCCTTTCC
7	1	BIM-3D	GGAAAGGAAATAAATAGAAATTTAATTG	BIM-3R	AATTACAAAAAATAACCTAAAAAAACTC
	2	BIM-4D	ATTGTAGGAGGTGGAATTTATAATT	BIM-4R	CCAACCTCCATCTCCTCAAAAC
8	1	BIM-5AD	GAGGGATTTGTTTTAATAATGTTTTG	BIM-5AR	CACCCACACTCTACCCCTACT
	2	BIM-5D	GAAGGTAGTTAGTTTAGAGGAGG	BIM-5R	CAACAATTATAACACAAACTAATTCTT
	3	BIM-6D	GTTTTGAGGAGATAGGAGGTTGG	BIM-6R	CATCACCAAAACAAATTAACATACC
9	1	BIM-7D	GGTATGTTAATTTGTTTTGGTGATG	BIM-7R	AAACTACCAACACACAAATATAAATAC
	2	BIM-8D	AAGAATTAGTTTGTGTTATAATTGTTG	BIM-8R	TATCAAACAACCCCAATTCACTAC
10	1	BIM-21D	TGGGAATTTTGTAGATTAGGTTGG	BIM-21R	TAAAAATCACTCTTATTCAAACTACC
	2	BIM-22D	AAAGTAAAAGGTTTTTGTTATTAAGGT	BIM-22R	TTTTCCCAATAAAAACAAAAACCTC
11	1	BIM-19R	TTTTAAGTTTGTTTGGGTATTATTATG	BIM-19R	CATAAATTCATTAAATATACCACCATC
	2	BIM-20R	TTATTATGTTGATGATGGATGGATTA	BIM-20R	AAAAAAAATTCTACCAAAATCTTCCC
12	1	BIM-17D	GTGATTTGGGGAATAGGTGGAG	BIM-17R	AAAAACAAACTACTACATACTAAACC
	2	BIM-18D	TTAGTTTGGATAGTATGTTATTGGA	BIM-18R	CCCTTTTCAAATAAACATACAAAC
13	1	BIM-15D	TGTTTGATGAGAAAGAATTAAATAGG	BIM-15R	CCTAAATTCAATCCCCAAAAATAC
	2	BIM-16D	GTATTTTTTAAGATTTGTTTGTGTTTG	BIM-16R	CATCTCCCTACCCCTACCCAC
14	1	BIM-9D	GTGTTTAGTAGTGGGTTTTATTTTAT	BIM-9D	AAAACTAAAACCAAACCTCCTACC
	2	BIM-10D	TAAAAAGAATTAAAGTAGGTATGATGA	BIM-10R	TACCAATAAAATCCAACTCAAAACTA
15	1	BIM-13D	GAAAGATTGTTATATTTTGAGGTATG	BIM-13R	CCTCAAATCACATAAACACTTACT
	2	BIM-14D	TTGAAATAGATTTATTTTTGAGGTTTG	BIM-14R	CTCTCCCTCTACTTAATACCTTAA

In: Gene Silencing: New Research
Editor: Grace W. Redberry, pp. 143-163
ISBN 1-59454-832-3

Chapter V

Intron-Mediated Gene Silencing

Shi-Lung Lin* and Shao-Yao Ying
Department of Cell and Neurobiology, Keck School of Medicine, University of Southern California, Los Angeles, CA 90033

Abstract

Nearly 97% of the human genome is the non-coding DNA, which varies from one species to another, and changes in these sequences are frequently noticed to manifest clinical and circumstantial malfunction. Numerous non-protein-coding genes are recently found to encode microRNAs, which are responsible for RNA-mediated gene silencing through RNA interference (RNAi)-like pathways. MicroRNAs (miRNAs), small single-stranded regulatory RNAs capable of interfering with intracellular messenger RNAs (mRNAs) that contain either complete or partial complementarity, are useful for the design of new therapies against cancer polymorphism and viral mutation. This flexible characteristic is different from double-stranded siRNAs (small interfering RNAs) because a much more rigid complementarity is required for siRNA-induced RNAi gene silencing. MiRNAs were firstly discovered in *Caenorhabditis elegans* as native RNA fragments that modulate a wide range of genetic regulatory pathways during embryonic development. Currently, varieties of miRNAs are widely reported in plants, animals, and even microbes. Intronic microRNA is a new class of miRNAs derived from the processing of gene introns. The intronic miRNAs differ uniquely from previously described intergenic miRNAs in the requirement of type II RNA polymerases (Pol-II) and spliceosomal components for their biogenesis. Several kinds of Id-miRNAs have been identified in *C. elegans*, mouse and human cells; however, neither function nor application has been reported. Here, we show for the first time that intron-derived miRNAs are able to induce RNA interference in not only human and mouse cells but also zebrafishes, chicken embryos and adult mice, demonstrating the evolutionary preservation of the intron-mediated gene silencing through miRNA functionality *in cell* and *in vivo*. These findings suggest an intracellular miRNA-mediated gene regulatory system, fine-tuning the degradation of protein-coding messenger RNAs.

Introduction

Intron occupies the largest proportion of non-coding sequences in the protein-coding DNA of a genome. The transcription of the genomic protein-coding DNA generates precursor messenger RNA (pre-mRNA), which contains four major parts including 5'-untranslational region (UTR), protein-coding exon, non-coding intron and 3'-UTR. In broad definition, both 5'- and 3'-UTR can be seen as a kind of intron extension; however, their processing during mRNA translation is different from the intron located between two protein-coding exons, or termed the in-frame intron. The in-frame intron can be sized up to several ten kilo-base nucleotides and was thought to be a huge genetic waste in gene transcripts. Recently, this stereotype mis-understanding was changed after the finding of intron-derived microRNA (intronic miRNA). Intronic miRNA is a new class of small single-stranded regulatory RNAs derived from the processing of pre-mRNA introns. Approximately 10~30% of a spliced intron are exported into cytoplasm with a moderate half-life [1]. MiRNA is usually sized about 18–25 oligonucleotide capable of either directly degrading its intracellular messenger RNA (mRNA) target or suppressing the protein translation of its targeted mRNA, depending on the complementarity between the miRNA and its target. In this way, the intronic miRNA is similar to previously described intergenic miRNAs structurally and functionally, but differs from them in its unique requirement of Pol-II and RNA splicing components for biogenesis [2,3,4]. Several types of intronic miRNA have been identified in *C. elegans*, mouse and human genomes [3,5,6] and some of their functions have been related to RNA interference (RNAi).

RNA interference (RNAi) is a posttranscriptional gene silencing mechanism in eukaryotes, which can be triggered by small RNA molecules such as microRNA (miRNA) and small interfering RNA (siRNA). These small RNA molecules usually function as a gene silencer, interfering with intracellular expression of genes either completely or partially complementary to the small RNAs. In principle, siRNAs are double-stranded RNAs capable of degrading target gene transcripts with almost perfect complementarity [7,8]. Unlike the stringent complementarity of siRNAs to their RNA targets, miRNAs are single-stranded and able to pair with target RNAs that have partial complementarity to the miRNAs [9,10]. Numerous natural miRNAs have been found to derived from hairpin-like RNA precursors in almost all eukaryotes, including yeast (*Schizosaccharomyces pombe*), plant (*Arabidopsis spp.*), nematode (*Caenorhabditis elegans*), fly (*Drosophila melanogaster*), mouse and human, involving defense against viral infections and regulation of certain gene expressions during development [11–21]. In contrast, natural siRNAs were abundantly discovered in plants and low-level animals (worms and flies), but rarely in mammals [10]. Because of the widespread detection of miRNAs in eukaryotes, these small RNAs have recently been used to design novel therapeutics against cancers and viral infections [2,22]. In fact, gene silencing mechanisms involving miRNAs have been proposed to be an intracellular defense system for eliminating undesired transgenes and foreign RNAs, such as viral infections and retrotransposon activities [22,23].

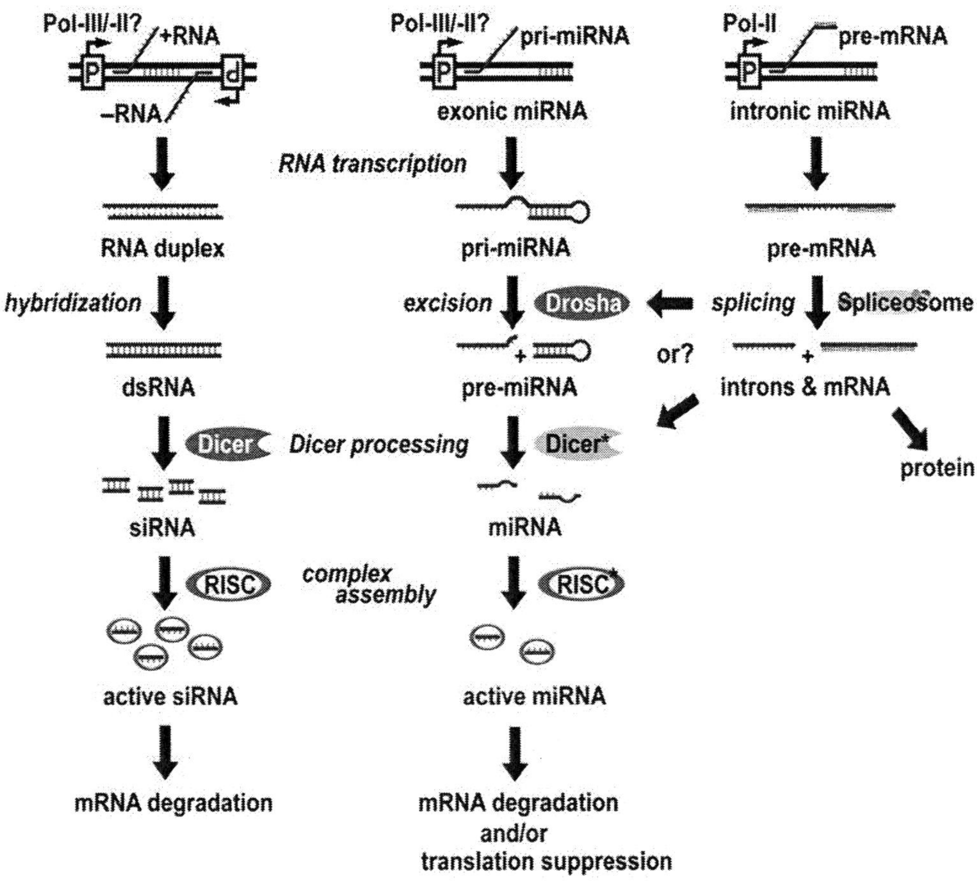

Figure 1: Comparison of biogenesis and RNAi mechanisms among siRNA, intergenic (exonic) miRNA and intronic miRNA. SiRNA is likely formed by two perfectly complementary RNAs transcribed from two different promoters (remains to be determined) and further processing into 19-22 bp duplexes by RNase III-familial endonuclease, Dicer. The biogenesis of intergenic miRNAs, *e.g. lin*-4 and *let*-7, involves a long transcript precursor (pri-miRNA), which is probably generated by Pol-II or Pol-III RNA promoters, while intronic miRNAs are transcribed by the Pol-II promoters of its encoded genes and co-expressed in the intron regions of the gene transcripts (pre-mRNA). After RNA splicing and further processing, the spliced intron may function as a pri-miRNA for intronic miRNA generation. In the nucleus, the pri-miRNA is excised by Drosha RNase to form a hairpin-like pre-miRNA template and then exported to cytoplasm for further processing by Dicer* to form mature miRNAs. The Dicers for siRNA and miRNA pathways are different. All three small regulatory RNAs are finally incorporated into a RNA-induced silencing complex (RISC), which contains either strand of siRNA or the single-strand of miRNA. The effect of miRNA is considered to be more specific and less adverse than that of siRNA because only one strand is involved. On the other hand, siRNAs primarily trigger mRNA degradation, whereas miRNAs can induce either mRNA degradation or suppression of protein synthesis depending on the sequence complementarity to the target gene transcripts.

The process of miRNA biogenesis in vertebrates presumably involves five steps (Figure 1). First, miRNA is generated as a long primary precursor miRNA (pri-miRNA) probably by RNA polymerases type II (Pol-II) [3,24]. Second, the long pri-miRNA is excised by Drosha-like RNase III endonucleases and/or spliceosomal components to form precursor miRNA (pre-miRNA), depending on the origin of the pri-miRNA located in an exon or an intron respectively [3,25], and third, the pre-miRNA is exported out of the nucleus by Ran-GTP and a receptor Exportin-5 [26,27]. In the cytoplasm, Dicer-like nucleases cleave the pre-miRNA to form mature miRNA. Lastly, the mature miRNA is incorporated into a ribonuclear particle (RNP), which becomes the RNA-induced gene silencing complex (RISC) capable of executing RNAi-related gene silencing effects [28,29]. Although the *in vitro* model of siRNA-associated RISC assembly has been reported, the link between the final miRNA maturation and RISC assembly remains to be determined. The characteristics of Dicer and RISC have been reported to be distinct between the siRNA and miRNA mechanisms [30]. In zebrafish, we have recently observed that the stem-loop structure of pre-miRNAs is involved in the strand selection for mature miRNA during RISC assembly. These findings suggest that the duplex structure of siRNA may be not essential for the assembly of miRNA-associated RISC *in vivo*. The biogenesis of miRNA and siRNA seems to share high similarity, however, the miRNA mechanisms previously proposed were based on the model of siRNA, and thus one must distinguish their individual properties and differences in these two types of RNAs to understand the evolutional and functional relationship of these gene silencing pathways. In addition, the differences may provide a clue for the prevalence of native siRNAs in invertebrates but rarely in mammals.

The definition of intronic miRNA is based on two factors; first, they must share the same promoter with their encoded genes, and second, they are spliced out of the transcript of their encoded genes and further processed into mature miRNAs. Although some of currently identified miRNAs are encoded in the genomic intron region of a gene but in the opposite orientation to the gene transcript, those miRNAs are not intronic miRNAs because they neither share the same promoter with the gene nor need to be released from the gene transcript by RNA splicing. For the transcription of those miRNAs, their promoters are located in the antisense direction to the gene, likely using the gene transcript as a potential target for the antisense miRNAs. A good example is *let-7c*, which was found to be an intergenic miRNA located in the antisense region of a gene intron. Current computer programs for miRNA prediction cannot distinguish the intronic miRNAs from the intergenic miRNAs. Because intronic miRNAs are encoded in the gene transcript precursors (pre-mRNAs) and share the same promoter with the encoded gene transcripts, the miRNA prediction programs tend to classify the intronic miRNAs as same as the intergenic miRNAs located in the exonic regions of genes. According to the differently biogenic mechanisms between the intronic and exonic miRNAs, these two types of miRNAs may function as different gene regulators in the adjustment of cellular physiology. Thus, a miRNA-prediction program including database of non-coding sequences located in the protein-coding pre-mRNA regions is urgently needed for thoroughly screening and understanding the distribution and variety of hairpin-like intronic miRNAs in the genomes.

Conclusion

The majority of human gene transcripts contain introns, which vary from species to species, and changes in these non-protein-coding sequences are frequently observed in clinical and circumstantial malfunction. Numerous introns are recently found to encode miRNAs, which are involved in RNAi-related chromatin silencing mechanisms. Over 90 intronic miRNAs have been identified using the bioinformatic approaches to date, but the function of the vast majority of these molecules remains to be determined [6]. According to the strictly expressive correlation of intronic miRNAs to their encoded genes, one may speculate that the levels of condition-specific, time-specific, and individual-specific gene expression are determined by interactions of distinctive miRNAs on single or multiple gene modulation. This interpretation accounts for more accurate genetic expression of various traits and any dysregulation of the interactions will thus result in genetic diseases. For instance, monozygotic twins frequently demonstrate slightly, but definitely distinguishing, disease susceptibility and physiological behavior. Such as, a long CCTG expansion in the intron 1 of a zinc finger protein ZNF9 gene has been correlated to type 2 myotonic dystrophy in either one of the twins with higher susceptibility [31]. Since the expansion motif obtained high affinity to certain RNA-binding proteins, the interfering role of intron-derived expansion fragments remains to be elucidated. Another more established example involving intronic expansion fragments in its pathogenesis is fragile X syndrome, which represents about 30% of human inherited mental retardation. Intronic CGG repeat (rCGG) expansion in the 5'-UTR of *FMR1* gene is the causative mutation in 99% of individuals with fragile X syndrome [32]. *FMR1* encodes an RNA-binding protein, FMRP, which is associated with polyribosome assembly in an RNP-dependent manner and capable of suppressing translation through an RNAi-like pathway. FMRP also contains a nuclear localization signal (NLS) and a nuclear export signal (NES) for shuttling certain mRNAs between the nucleus and cytoplasm [33]. Jin *et al* proposed an RNAi-mediated methylation model in the CpG region of *FMR1* rCGG expansion, which is targeted by a hairpin RNA derived from the 3'-UTR of the *FMR1* expanded allele transcript [32]. The Dicer-processed hairpin RNA triggers the formation of RNA-induced initiator of transcriptional gene silencing (RITS) on the homologous rCGG sequences and leads to heterochromatin repression of the *FMR1* locus. These examples suggest that natural evolution gives raise to more complexity and more variety of introns in higher animals and plants for coordinating their vast gene expression volumes and interactions; therefore, any dysregulation of miRNAs derivation from introns may thus lead to genetic diseases involving intronic expansion or deletion, such as myotonic dystrophy and fragile X mental retardation.

Vector-based Intronic miRNA Generation

To understand the disease caused by dysregulation of intronic miRNAs, an artificial expression system is needed to recreate the function and mechanism of the miRNA *in vitro* and *in vivo*. The same approach may be used to design and develop therapies for the disease. Previously, several vector-based RNAi expression systems have been developed, using type-

III RNA polymerase (Pol-III)-directed transcription activities, to generate more stable RNAi efficacy and lower interferon-related toxicity in several cell lines *in vitro* [34,35,36,37]. For gene therapy *in vivo*, a functional gene is preferably delivered into an animal or human being by expression-competent vector vehicles, such as retroviral vector, lentiviral vector, adenoviral vector, and adeno-associated viral (AAV) vector. The main purpose of these vector-based approaches is to maintain long-term and consistent gene modulation. Although some studies [38,39] attempting to use the Pol-III-directed RNAi system have succeeded in maintaining constant gene silencing efficacy *in vivo*, their delivery strategies failed to target a specific cell population due to the ubiquitous existence of Pol-III activity in all cell types. Moreover, the requirement of using Pol-III RNA promoters, *e.g.* U6 and H1, for small RNA expression causes another drawback. Because the read-through side-effect of Pol-III occurs on a short transcription template if without proper termination, large RNA products longer than desired 18-25 base-pairs (bp) can be synthesized and then cause unexpected interferon cytotoxicity [40,41]. Such a problem can also result from the competitive conflict between the Pol-III promoter and another vector promoter (*i.e.* LTR and CMV promoters). Sledz *et al.* and us have found that high dosage of siRNAs (e.g., > 250 nM in human T cells) was able to cause strong cytotoxicity similar to that of long double-stranded dsRNAs [42,43]. This toxicity is due to the double-stranded structure of siRNAs and dsRNAs, which activates the interferon-mediated non-specific RNA degradation and programmed cell death through the signaling of PKR and 2-5A systems. It is well known that interferon-induced protein kinase PKR can trigger cell apoptosis, while activation of interferon-induced 2',5'-oligoadenylate synthetase (2-5A) system leads to extensive cleavage of single-stranded RNAs (*i.e.* mRNAs) [44]. Both PKR and 2-5A systems contain dsRNA-binding motifs which are highly conserved for binding to dsRNAs, but these motifs do not bind to either single-strand RNAs or RNA-DNA hybrids. These disadvantages limit the use of Pol-III-based RNAi vector systems for the therapeutical purpose.

The intron-derived miRNA system is able to be activated in a specific cell type under the control of a type-II RNA polymerases (Pol-II)-directed transcriptional machinery. To overcome the Pol-III-mediated siRNA side-effects, we have successfully developed a novel Pol-II-based miRNA biogenesis strategy, employing intronic miRNA molecules [3] to knock down more than 85% of selected oncogene function or viral genome replication [45,46]. Because of the flexibility in binding with partially complementary mRNA targets, miRNA can serve as an anti-cancer drug or vaccine to achieve a major breakthrough in the treatments of cancer polymorphisms and viral mutations. We are the first research group who discovered the biogenesis of miRNA-like precursors from the 5'-proximal intron regions of gene transcripts (pre-mRNAs) produced by the mammalian Pol-II. Depending on the promoter of the miRNA-encoded gene transcript, intronic miRNA is co-expressed with its encoding gene in the specific cell population, which activates the promoter and expresses the gene. It has been noted that a spliced intron was not completely digested into monoribonucleotides for transcriptional recycling since approximately 10–30% of the intron was found in the cytoplasm with a moderate half-life [1,47]. This type of miRNA generation has been found to rely on the coupled interaction of nascent Pol-II-mediated pre-mRNA transcription and intron excision, occurring within certain nuclear regions proximal to genomic perichromatin fibrils [46,48,49]. After Pol-II RNA processing and splicing excision, some of the intron-derived miRNA fragments can form mature miRNAs and effectively silence the target genes through the RNAi mechanism, while the exons of pre-mRNA are ligated together to form a mature mRNA for protein synthesis (Figure 2A) [3]. Because miRNAs are single-stranded molecules insensitive to PKR- and 2-5A-induced interferon systems, the utilization of this Pol-II-mediated miRNA generation can be safe *in vitro and in vivo* without the cytotoxic effects of dsRNAs and siRNAs. These findings indicate new functions for mammalian introns in intracellular miRNA generation and gene silencing, which can be used as a tool for analysis of gene functions and development of gene-specific therapeutics against cancers and viral infections.

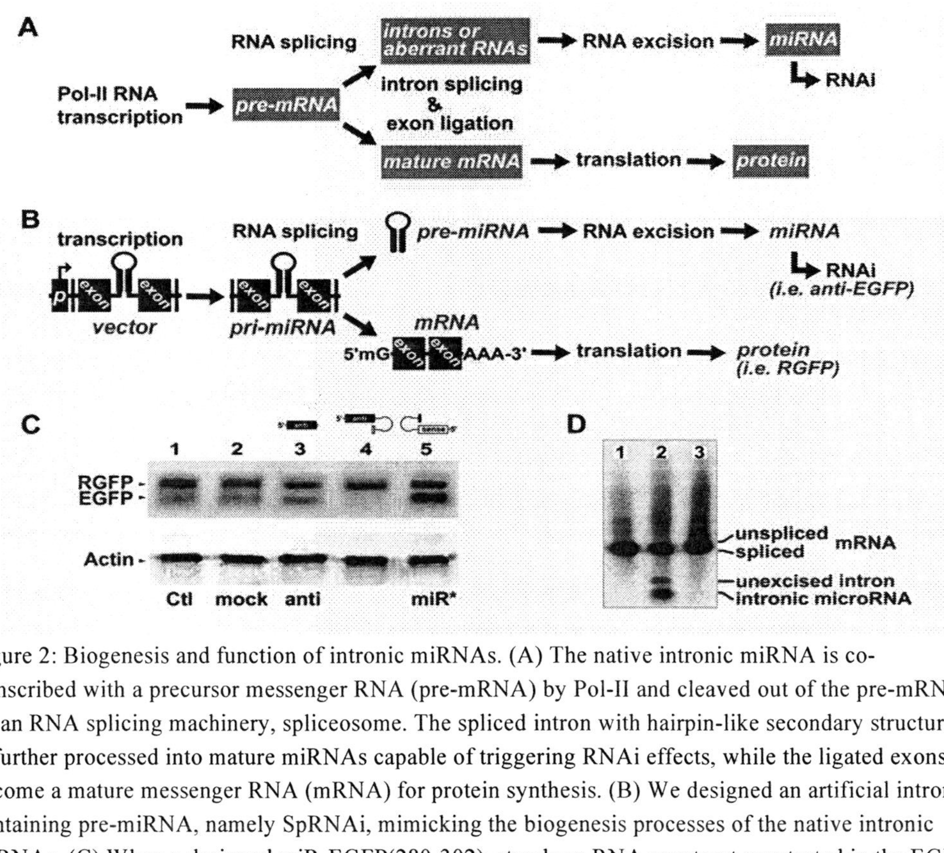

Figure 2: Biogenesis and function of intronic miRNAs. (A) The native intronic miRNA is co-transcribed with a precursor messenger RNA (pre-mRNA) by Pol-II and cleaved out of the pre-mRNA by an RNA splicing machinery, spliceosome. The spliced intron with hairpin-like secondary structures is further processed into mature miRNAs capable of triggering RNAi effects, while the ligated exons become a mature messenger RNA (mRNA) for protein synthesis. (B) We designed an artificial intron containing pre-miRNA, namely SpRNAi, mimicking the biogenesis processes of the native intronic miRNAs. (C) When a designed miR-EGFP(280-302)–stemloop RNA construct was tested in the EGFP-expressing Tg(UAS:gfp) zebrafishes, we detected a strong RNAi effect only on the target EGFP (lane 4). No detectable gene silencing effect was observed in other lanes from left to right: 1, blank vector control (Ctl); 2, miRNA–stemloop targeting HIV-p24 (mock); 3, miRNA without stemloop (anti); and 5, stemloop–miRNA* complementary to the miR-EGFP(280-302) sequence (miR*). The off-target genes such as vector RGFP and fish actin were not affected, indicating the high target specificity of miRNA-mediated gene silencing. (D) Three different miR-EGFP(280-302) expression systems were tested for miRNA biogenesis from left to right: 1, vector expressing intron-free RGFP, no pre-miRNA insert; 2, vector expressing RGFP with an intronic 5'-miRNA-stemloop-miRNA*-3' insert; and 3, vector similar to the 2 construct but with a defected 5'-splice site in the intron. In Northern bolt analysis probing the miR-EGFP(280-302) sequence, the mature miRNA was released only from the spliced intron resulted from the vector 2 construct in the cell cytoplasm.

Using artificial introns carrying hairpin-like miRNA precursors (pre-miRNA), we have successfully generated mature miRNA molecules with full capacity in triggering RNAi-like gene silencing in human prostate cancer LNCaP, human cervical cancer HeLa and rat neuronal stem HCN-A94-2 cells [3, 45]. As shown in Figure 2B, the artificial intron (SpRNAi) was co-transcribed within a precursor messenger RNA (pre-mRNA) by Pol-II and cleaved out of the pre-mRNA by RNA splicing. Then, the spliced intron containing a pre-miRNA structure was further processed into mature miRNAs capable of triggering RNAi-related gene silencing effects. Based on this artificial miRNA model, we have tested various pre-miRNA constructs, and observed that the production of intron-derived miRNA fragments was originated from the 5'-proximity of the intron sequence between the 5'-splice site and the branching point. These miRNAs were able to trigger strong suppression of genes possessing more than 70% of complementarity to the miRNA sequences, whereas non-homologous miRNAs, *i.e.* empty intron without the pre-miRNA insert, intron with an off-target miRNA insert (negative control), and splicing-defective intron showed no silencing effects on the targeted gene. The same results can also be reproduced in the zebrafish directed against target EGFP expression (Figure 2C), indicating the consistent preservation of the intronic miRNA biogenesis system in vertebrates. Further, no effect was detected on off-target genes, such as RGFP and β-actin, suggesting the high specificity of miRNA-directed RNA interference (RNAi). We have confirmed the identity of the intron-derived miRNAs, which were sized about 18–25 base nucleotides (nt), approximately similar to the newly identified intronic miRNAs in *C. elegans*. Moreover, the intronic small RNAs isolated by guanidinium-chloride ultracentrifugation can elicit strong, but short-term gene silencing effects on the homologous genes in transfected cells, indicating their temporary RNAi effects. Thus, the long-term (> 1 month) gene silencing effect that we observed *in vivo*, using the Pol-II-mediated intronic miRNA system, is likely maintained by constitutively miRNA production from the vector rather than the stability of the miRNAs.

The components of the Pol-II-mediated SpRNAi system include several consensus nucleotide elements consisting of a 5'-splice site, a branch-point domain, a poly-pyrimidine tract and a 3'-splice site (Figure 3). Additionally, a pre-miRNA insert sequence is placed within the artificial intron between the 5'-splice site and the branch-point domain. This portion of the intron would normally form a lariat structure during RNA splicing and processing. We currently know that spliceosomal U2 and U6 snRNPs, both helicases, may be involved in the unwinding and excision of the lariat RNA fragment into pre-miRNA; however, the detailed processing remains to be elucidated. Further, the SpRNAi contains a translation stop codon domain (T codon) in its 3'-proximal region to facilitate the accuracy of RNA splicing, which if present in a cytoplasmic mRNA, would signal the diversion of a splicing-defective pre-mRNA to the nonsense-mediated decay (NMD) pathway and thus cause the elimination of any unspliced pre-mRNA in the cell. For intracellular expression of the SpRNAi, we need to insert the SpRNAi construct into the *Dra*II cleavage site of a red fluorescent membrane protein (RGFP) gene from mutated chromoproteins of coral reef *Heteractis crispa*. The cleavage of RGFP at its 208th nucleotide site by restriction enzyme *Dra*II generates an AG GN nucleotide break with three recessing nucleotides in each end, which will forms 5' and 3' splice site respectively after the SpRNAi insertion. Because this intronic insertion disrupts the expression of functional RGFP, it becomes possible to determine the occurrence of intron splicing and RGFP-mRNA maturation through the appearance of red fluorescent emission around the membrane surface of the transfected cells. The RGFP also provides multiple exonic splicing enhancers (ESEs) to increase RNA splicing efficiency.

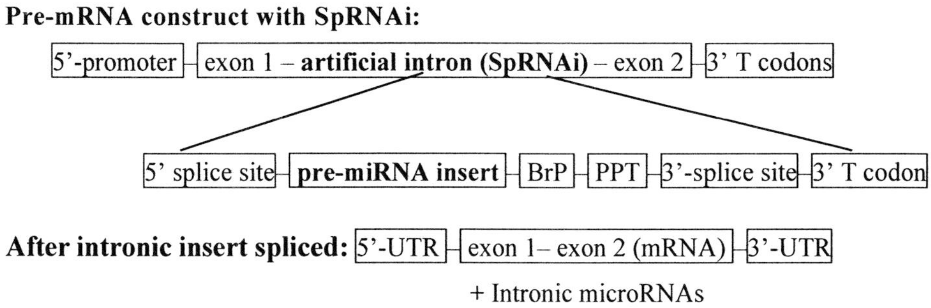

Figure 3: Schematic construct of the artificial SpRNAi intron in a recombinant gene SpRNAi-RGFP for intracellular expression and processing. The components of the Pol-II-mediated SpRNAi system include several consensus nucleotide elements consisting of a 5'-splice site, a branch-point domain (BrP), a poly-pyrimidine tract (PPT), a 3'-splice site and a pre-miRNA insert located between the 5'-splice site and the BrP domain. The expression of the recombinant gene is under the regulation of either a mammalian Pol-II RNA promoter or a compatible viral promoter for cell-type-specific effectiveness. Mature miRNAs are released from the intron by RNA splicing and further Dicer processing.

Intron-mediated Gene Silencing in Zebrafishes

The foregoing establishes the fact that intronic miRNAs can be used as an effective strategy to silence specific target gene *in vivo*. We firstly tried to resolve the structural design of pre-miRNA inserts for the best gene silencing effect and found out that a strong structural bias exists in the selection of a mature miRNA strand during assembly of the RNAi effector, RNA-induced gene silencing complex (RISC). RISC is a protein–RNA complex that directs either target gene transcript degradation or translational repression through the RNAi mechanism. Formation of siRNA duplexes has been reported to play a key role in assembly of the siRNA-associated RISC. The two strands of the siRNA duplex are functionally asymmetric, but assembly into the RISC complex is preferential for only one strand. Such preference is determined by the thermodynamic stability of each 5'-end base-pairing in the strand. Based on this siRNA model, the formation of miRNA and its complementary miRNA (miRNA*) duplexes was thought to be an essential step for the assembly of miRNA-associated RISC. If this were true, no functional bias would be observed in the stemloop of a pre-miRNA. Nevertheless, we observed that the stemloop of the intronic pre-miRNA was involved in the strand selection of a mature miRNA for RISC assembly in zebrafish. In these experiments, we constructed miRNA-expressing SpRNAi-RGFP vectors as previously described [3] and two symmetric pre-miRNAs, miRNA-stemloop-miRNA* (❶) and miRNA*-stemloop-miRNA (❷), were synthesized and inserted into the vectors, respectively. Both pre-miRNAs contained the same double-stranded stem arm region, which was directed against the EGFP nts 280–302 sequence. Because the intronic insert region of the SpRNAi-

RGFP recombined gene is flanked with a *PvuI* and an *MluI* restriction site at the 5'- and 3'-ends, respectively, the primary insert can be easily removed and replaced by various gene-specific inserts (e.g. anti-EGFP) possessing cohesive ends. By changing the pre-miRNA inserts directed against different gene transcripts, this intronic miRNA generation system provides a valuable tool for genetic and miRNA-associated research *in vivo*.

To determine the structural preference of the designed pre-miRNAs, we have isolated the zebrafish small RNAs by mirVana miRNA isolation columns (Ambion, Austin, TX) and then precipitated all potential miRNAs complementary to the target EGFP region by latex beads containing the target RNA sequence. One full-length miRNA identity, miR-EGFP(280-302), was verified to be active in the transfections of the 5'-miRNA-stemloop-miRNA*-3' construct, as shown in the Figure 4A (gray-shading sequences). Since the mature miRNA was detected only in the zebrafish transfected by the 5'-miRNA-stemloop-miRNA*-3' construct, the miRNA-associated RISC tends to preferably interact with the construct ❷ rather than the ❶ pre-miRNA. The green fluorescent protein EGFP expression was constitutively driven by the β-actin promoter located in almost all cell types of the zebrafish, while the Figure 4B shows that transfection of the SpRNAi-RGFP vector into the Tg(UAS:gfp) zebrafish co-expressed the red fluorescent protein RGFP, serving as a positive indicator for the miRNA generation in the transfected cells. This approach has been successfully used in several mouse and human cell lines to show the RNAi effects [3,45]. We applied the liposome-capsulated vector (total 60 μg) to the fishes and found that the vector easily penetrated almost all tissues of the two-week-old zebrafish larvae within 24 hours, reaching fully systemic delivery of the miRNA effect. The indicator RGFP was detected in both of the fishes transfected by either 5'-miRNA*-stemloop-miRNA-3' or 5'-miRNA-stemloop-miRNA*-3' pre-miRNA, whereas the silencing of target EGFP expression (green) was observed only in the fish transfected by the 5'-miRNA-stemloop-miRNA*-3' pre-miRNA (Figures 1B and 1C). The suppression level in the gastrointestinal (GI) tract was found to be less effective, probably due to the high RNase activity in this region. Because thermostability in the 5' end of the siRNA duplexes resulting from both of the designed pre-miRNAs is the same, we suggest that the stemloop of pre-miRNA is involved in strand selection of mature miRNA during RISC assembly. Given that the cleavage site of Dicer in the stemarm determines the strand selection of mature miRNA [25], the stemloop may function as a determinant for the recognition of a special cleavage site. Therefore, the different stemloop structures among various species may also provide a clue for the prevalence of native siRNAs in invertebrates but rarely in mammals.

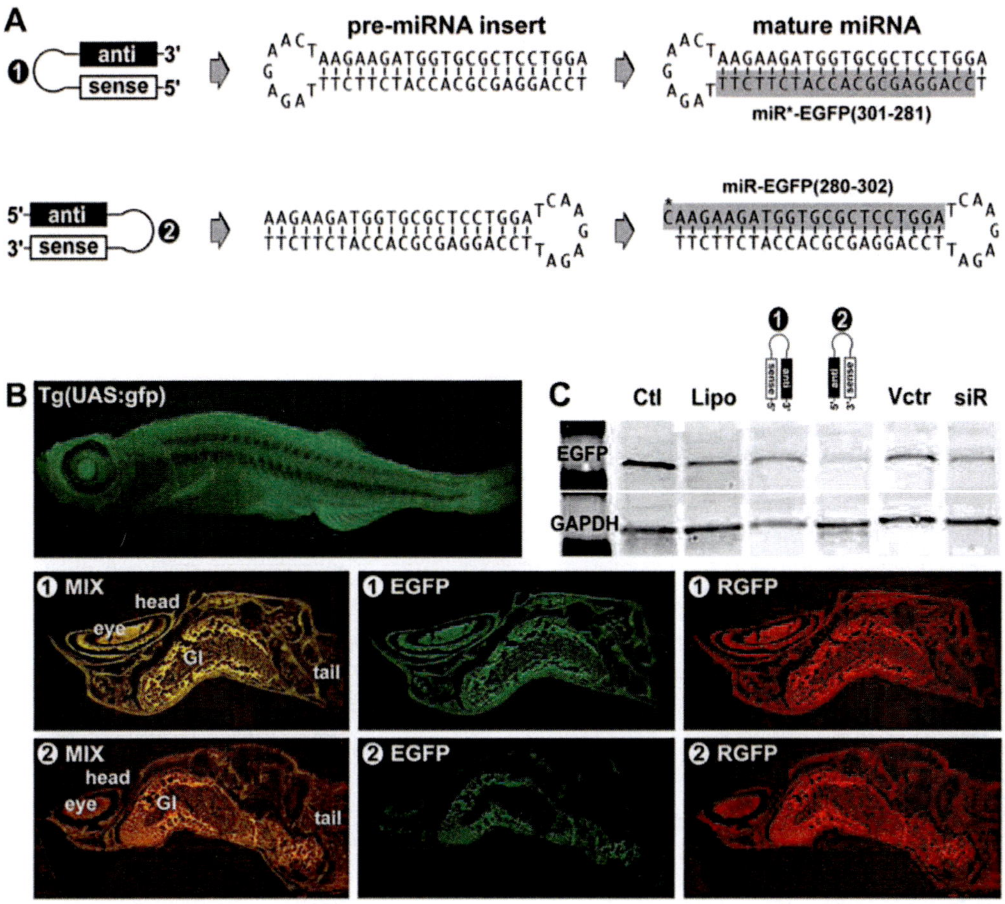

Figure 4: Bias of miRNA–miRNA* asymmetry in RISC *in vivo*. Different preferences of RISC assembly were observed by transfection of 5'-miRNA*-stemloop-miRNA-3' (❶) and 5'-miRNA-stemloop-miRNA*-3' (❷) pre-miRNA constructs in zebrafish, respectively. (A) Based on the assembly rule of siRNA RISC, the processing of both ❶ and ❷ should result in the same siRNA duplex for RISC assembly; however, the experiments demonstrate that only the ❷ construct was used in RISC assembly for silencing target EGFR. Because the miRNA is predicted to be complementary to its target messenger RNA, the "antisense" (black bar) refers to the miRNA and the "sense" (white bar) refers to its complementarity, miRNA*. One mature miRNA, namely miR-EGFP(280/302), was detected in the ❷-transfected zebrafishes, whereas the ❶ transfection produced another kind of miRNA, miR*-EGFR(301-281), which was partially complementary to the miR-EGFP(280/302). (B) *In vivo* gene silencing efficacy was only observed in the transfection of the ❷ pre-miRNA construct, but not the ❶ construct. Since the color combination of EGFP and RGFP displayed more red than green (as shown in deep orange), the expression level of target EGFP (green) was significantly reduced in ❷, while miRNA indicator RGFP (red) was evenly present in all vector transfections. (C) Western blot analysis of the EGFP protein levels confirmed the specific silencing result of (B). No detectable gene silencing was observed in fishes without (Ctl) and with liposome only (Lipo) treatments. The transfection of either a U6-driven siRNA vector (siR) or an empty vector (Vctr) without the designed pre-miRNA insert resulted in no gene silencing significance.

Intron-mediated Gene Silencing in Chicken Embryos

The *in vivo* model of chicken embryos has been widely utilized in the research of developmental biology, signal transduction and flu vaccine development. We have successfully tested the feasibility of localized gene silencing *in vivo* by the intronic miRNA approach and also discovered that the interaction between pre-mRNA and genomic DNA may be essential for the miRNA biogenesis. As an example, the β-catenin gene was selected because its products play a critical role in the biological development and ontogenesis [50]. The β-catenin is known to be involved in the growth control of skin and liver tissues in chicken embryos. The loss-of-function of β-catenin is lethal in transgenic animals. As shown in Figure 5, experimental results demonstrated that the miRNAs derived from a long mRNA–DNA hybrid construct (≥ 150 bp) were capable of inhibiting β-catenin gene expression in the liver and skin of developing chicken embryos. This is due to the mechanism by which homologous recombination between the intronic miRNA and genomic DNA may account for a part of its specific gene silencing effect [46]. We have demonstrated that the [P^{32}]-labeled DNA component of an mRNA–DNA duplex construct in cell nuclear lysates was intact during the effective period of miRNA-induced RNA interference (RNAi) phenomena, while the labeled RNA part was replaced by cold homologues and excised into small RNA fragments within a 3-day incubation period. Since the intronic miRNA generation relies on a coupled interaction of nascent Pol-II-directed pre-mRNA transcription and intron excision occurring proximal to genomic perichromatin fibrils, the above observation indicates that the pre-mRNA–genomic DNA recombination may facilitate new miRNA generation by the Pol-II RNA excision for relatively long-term gene silencing. Alternatively, Pol-II may function as an RNA-dependent RNA polymerase (RdRp) for producing more small interfering RNAs, because mammalian Pol-II has been reported to possess RdRp activities [51,52]. Thus, it suggests that Pol-II-mediated RNA generation and excision is involved in both mRNA-DNA-derived and intron-derived miRNA biogenesis, resulting in single-stranded small RNAs of about 20 nt comparable to the general sizes of Dicer-processed miRNAs frequently observed in the regulations of numerous developmental events.

In an effort to examine the pre-mRNA and genome interaction theory, we tested intracellular transfection of the mRNA–DNA hybrid construct containing a hairpin anti-β-catenin pre-miRNA, which was directed against the central region of the β-catenin coding sequence (aa 306–644) with perfect complementarity. A perfectly complementary miRNA theoretically directs target mRNA degradation more efficient than translational repression. Using embryonic day 3 chicken embryos, a dose of 25 nM of the pre-miRNA construct was injected into the ventral body cavity, which is close to where the liver primordia would form (Figure 5A). For efficient delivery into target tissues, the pre-miRNA construct was mixed with the DOTAP® liposomal transfection reagent (Roche Biomedicals, Indianapolis, IN) at a ratio of 3:2. A 10% (v/v) fast green solution was concurrently added during the injection as a dye indicator. The mixtures were injected into the ventral side near the liver primordia below the heart using heat pulled capillary needles. After injection, the embryonic eggs were sealed with sterilized scotch tapes and incubated in a humidified incubator at 39-40°C until day 12 when the embryos were examined and photographed under a dissection microscope. Several malformations were observed, while the embryos still survived and there was no visible overt toxicity or overall perturbation of embryo development. The liver was the closest organ to the injection site and thus was most dramatically affected in its phenotypes. Other regions, particularly the skin close to the injection site, were also affected by the diffused miRNA effects. As shown in Figure 5B, Northern blot analysis detecting the target β-catenin mRNA expression in the dissected livers showed that β-catenin expression in the wild-type livers remained normal (lanes 1-3), whereas those of the miRNA-treated samples was decreased dramatically (lanes 4-6). The miRNA silencing effect degraded more then 98% of β-catenin mRNA expression in the embryonic chicken, but has no influence in the house-keeping gene GAPDH expression, indicating its high target specificity and very limited interferon-related cytotoxicity *in vivo*.

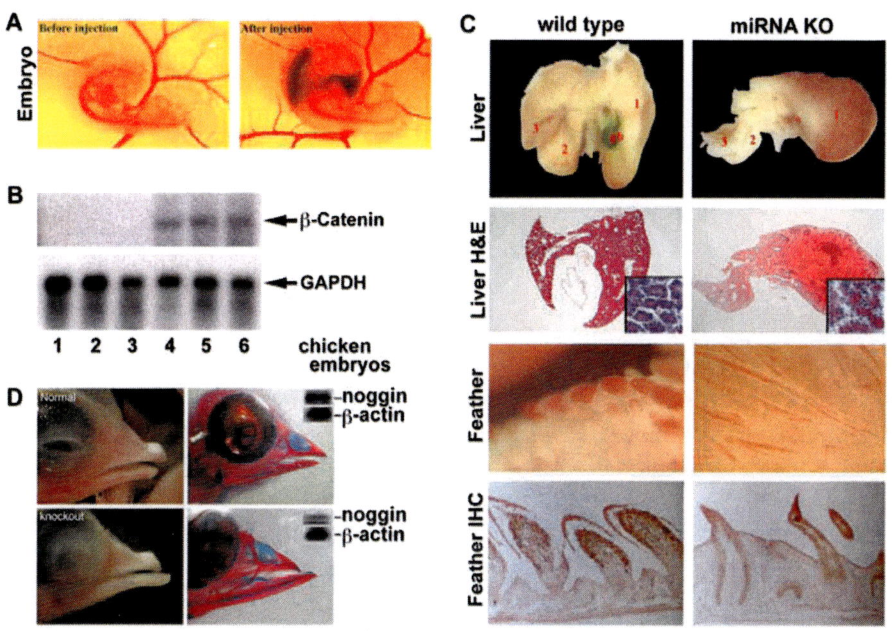

Figure 5: *In vivo* gene silencing effects of anti-β-catenin miRNA (A)–(C) and anti-noggin miRNA (D) on local organ development in embryonic chicken. (A) The pre-miRNA construct and fast green dye mixtures were injected into the ventral side of chicken embryos near the liver primordia below the heart. (B) Northern blots of extracted RNAs from chicken embryonic livers with (lanes 1, 2 and 3) and without (lanes 4, 5 and 6) anti-β-catenin miRNA treatments were shown. All three knockouts showed a more than 98% silencing effect on β-catenin mRNA expression but house-keeping genes, such as GAPDH, was not affected. (C) Liver formation of the β-catenin knockouts was significantly hindered (upper right 2 panels). Microscopic examination revealed a loose structure of hepatocytes, indicating the loss of cell-cell adhesion due to breaks in adherins junctions formed between β-catenin and cell membrane E-cadherin in early liver development. In severely affected regions, feather growth in the skin close to the injection area was also inhibited (lower right 2 panels). Immunohistochemistry staining of β-catenin protein expression (brown) showed a significant decrease in the feather follicle sheaths. (D) The lower beak development was increased by the mandible injection of the anti-noggin pre-miRNA construct (down panel) in comparison to the wild type (up panel). Right panels showed bone (alizarin red) and cartilage (alcian blue) staining to demonstrate the out growth of bone tissues in the lower beak of the noggin knockout. Northern blot analysis (small windows) confirmed a ~60% decrease of noggin mRNA expression in the lower beak area.

After ten days of primordial injection with the anti-β-catenin pre-miRNA template, the embryonic chicken livers showed an enlarged and engorged first lobe, but the size of the second and third lobes of the livers were dramatically decreased (Figure 5C). Histological sections of normal livers showed hepatic cords and sinusoidal space with few blood cells. In the anti-β-catenin miRNA-treated embryos, the general architecture of the hepatic cells in lobes 2 and 3 remained unchanged; however, there were islands of abnormal regions in lobe 1. The endothelium development appeared to be defective and blood leaked outside of the blood vessels. Abnormal types of hematopoietic cells were also observed between the space of hepatocytes, particularly dominated by a population of small cells with round nuclei and scanty cytoplasm. In severely affected regions, hepatocytes were disrupted (Figure 5C, small windows) and the diffused miRNA effect further inhibited the feather growth in the skin area close to the injection site. The results discussed above showed that the anti-β-catenin miRNA was very effective in knocking out the targeted gene expression at a very low dose of 25 nM and was effect over a long period of time ($\geq$ 10 days). Further, the miRNA gene silencing effect appeared to be very specific as off-targeted organs appear to be normal, indicating that the small single-strand composition of miRNA herein possessed no overt toxicity. In another attempt to silencing noggin expression in the mandible beak area using the same approach (Figure 5D), it was observed that an enlarged lower beak morphology is reminiscent of those of BMP4-overexpressing chicken embryos reported previously [53,54]. Skeleton staining showed the outgrowth of bone and cartilage tissues in the injected mandible area (Figure 5D, right panels) and Northern blot analysis further confirmed that about 60% of noggin mRNA expression was knocked out in this region (small windows). Since Bone morphogenetic protein 4 (BMP4), a member of transforming growth factor-β (TGF-β) superfamily, is known to promote bone development and the noggin is an antagonist of BMP2/4/7 genes, it is not surprised to find out that our miRNA-mediated noggin knockouts created a morphological change resemble to the BMP4-overexpressing results as previous reports in chicken and other avian models. Thus, the gene silencing in chicken by the pre-miRNA transfection presents a great potential of localized transgene-like approach in creating animal models for developmental biology research.

Intron-mediated Gene Silencing in Mouse Skins

To test the intronic miRNA effect on adult mammals (Figure 6), we used the vector-based miRNA delivery approach as previously reported in zebrafish. Patched albino (white) skins of melanin-knockout mice (W-9 black) were created by a succession of intra-cutaneous (*i.c.*) transduction of about anti-tyrosinase (Tyr) pre-miRNA construct (50 μg) for 4 days (total 200 μg). The Tyr, a type-I membrane protein and copper-containing enzyme, catalyzes the critical and rate-limiting step of tyrosine hydroxylation in the biosynthesis of melanin (black pigment) in skins and hairs. After 13-day incubation, the expression of melanin has been blocked due to a loss of its intermediates resulted from the anti-Tyr miRNA silencing effect. Contrarily, the blank control and the U6-directed siRNA-transfected mice presented normal skin color (black), indicating that the loss of melanin is specifically effective in the miRNA transfections. Moreover, Northern blot analysis using RNA-PCR-amplified mRNAs from hair follicles showed a 76.1±5.3% reduction of Tyr expression 2-day posterior to the miRNA transfection, in consistent with the immunohistochemistry results from the same skin area, whereas mild, non-specific degradation of common gene transcripts was detected in the siRNA-transfected skins (seen from smearing patterns of both house-keeping control GAPDH and target Tyr mRNAs). Thus, the results show that utilization of intronic miRNA vectors provides a powerful new strategy for *in-vivo* gene therapy, potentially to melanoma. Under the same dosage, the miRNA transfections did not cause detectable cytotoxicity effect, while the siRNA transfections induced noted non-specific mRNA degradation as previous reports [42,43]. This even underscores the fact that the miRNA is effective even under *in vivo* systems without the side-effects of dsRNA. The results also indicate that this gene silencing effect is stable and efficient in knocking down the target gene expression over a relatively long period of time since the hair regrowth requires at least ten-day recovery. Further, it was observed that non-targeted skin hairs appear to be normal, which implies that the intronic miRNA composition herein possess high specificity and no overt toxicity. Thus, the intronic miRNA approach offers relatively long-term, effective and safe gene manipulation in local animal tissues or organs, preventing the lethal effects of certain genes in the conventional transgenic animal model.

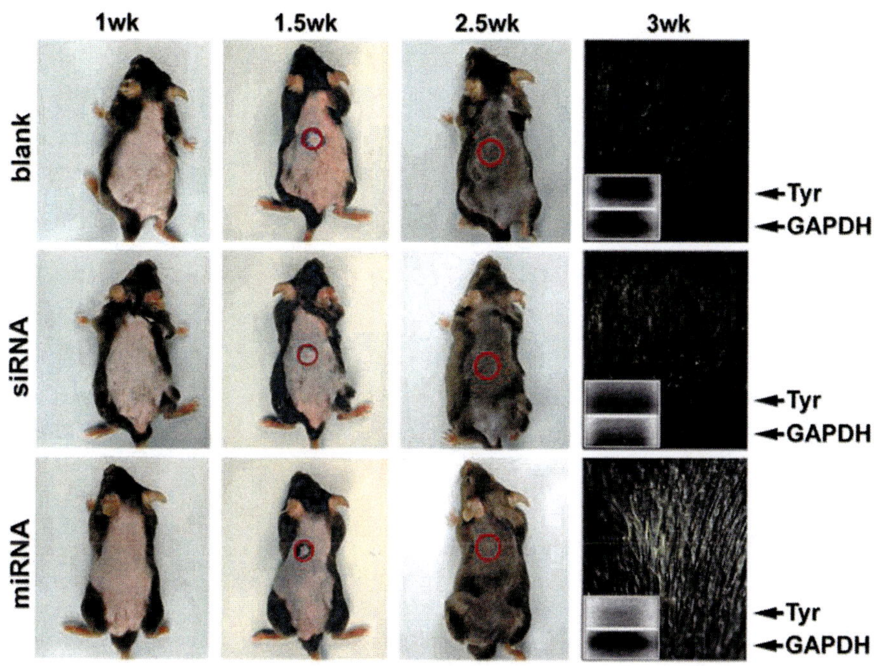

Figure 6: *In vivo* effects of anti-tyrosinase miRNA on the mouse pigment production of local skins. Transfection of the miRNA induced strong gene silencing of tyrosinase (Tyr) mRNA expression but not house-keeping GAPDH, whereas that of U6-directed siRNA triggered mild non-specific RNA degradation of both Tyr and GAPDH gene transcripts. Since Tyr is an essential enzyme for black pigment melanin production, the success of gene silencing can be observed by a significant loss of the black color in mouse hairs. The red circles indicate the location of *i.c.* injections. Northern blot analysis of Tyr mRNA expression in local hair follicles confirmed the effectiveness and specificity of the miRNA-mediated gene silencing effect (small windows).

Ex vivo anti-HIV Therapy Using Intronic miRNA Directed against Viral Genes

The following experimentations have shown the preliminary success of silencing exogenous retrovirus replication in an *ex vivo* cell model of patient-extracted $CD4^+$ T lymphocytes. The specific anti-HIV SpRNAi-RGFP vectors were designed to target against the *gag-pol* region of about the nts +2113 to +2450 of HIV-1 genome. This region is relatively conserved and can serve as a good target for anti-HIV treatment [55]. The viral genes located in this target region include 3'-proximal $Pr55^{gag}$ polyprotein (*i.e.*, matrix p17 + capsid p24 + nucleocapsid p7) and 5'-proximal $p66/p51^{pol}$ polyprotein (*i.e.*, protease p10 + reverse transcriptase); all these components have critical roles in viral replication and infectivity. During the early infection phase, the viral reverse transcriptase transcribes the HIV RNA genome into a double-stranded cDNA sequence, which forms a pre-integration complex with the matrix, integrase and viral protein R (Vpr). This complex is then transferred to the cell nucleus and integrated into the host chromosome, consequently establishing the HIV provirus. We hypothesized that, although HIV carries few reverse transcriptase and matrix proteins during its first entry into host cells, the co-suppression of $Pr55^{gag}$ and $p66/p51^{pol}$ gene expressions by miRNAs is expected to eliminate the production of infectious viral particles in the late infection phase. Silencing $Pr55^{gag}$ may prevent the assembly of intact viral particles due to the lack of matrix and capsid proteins, while suppression of protease in $p66/p51^{pol}$ can inhibit the maturation of several viral proteins. HIV expresses about nine viral gene transcripts which encode at least 15 various proteins, thus, the separation of a polyprotein into individual functional proteins requires the viral protease activity. As shown in Figure 7, this therapeutic approach has been reported to be feasible [22,43].

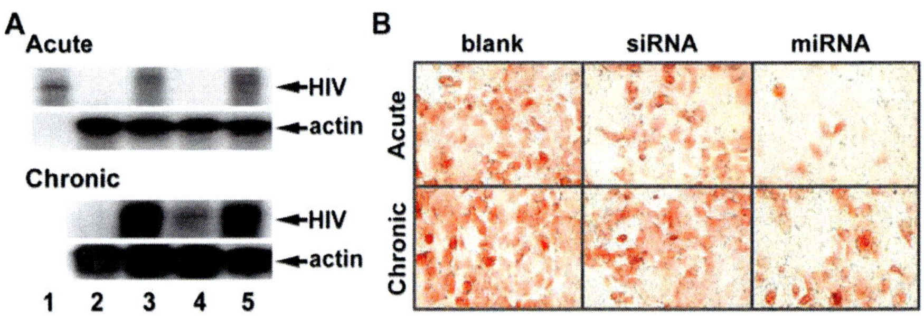

Figure 7: Silencing of HIV-1 genome replication using anti-*gag/pro/pol* miRNA transfections into $CD4^+$ T lymphocytes isolated from the acute and chronic phases of AIDS infections. (A) Northern blot analysis showed about 98% and 70% decreases of HIV genome in the acute and chronic infections after miRNA treatments (lanes 4), respectively. No effect was detected in the T cells transfected by miRNA* targeting the same *gag/pro/pol* region of the viral genome (lane 5). The size of pure HIV-1 provirus was measured about 9,700 nucleotide bases (lanes 1). RNA extracts from normal non-infected $CD4^+$ Th lymphocytes were used as a negative control (lanes 2), whereas those from HIV-infected T cells were used as a positive control (lanes 3). (B) Immunostaining of HIV p24 marker confirmed the results of (A). Since the *ex vivo* HIV-silenced T lymphocytes were resistant to any further infection by the same strains of HIV, they may be transfused back to the donor patient for eliminating HIV-infected cells.

The anti-HIV SpRNAi-RGFP vectors were tested in the $CD4^+$ T lymphocyte cells from HAART-treated, HIV-seropositive patients. Because only partial complementarity between miRNA and its target RNA is needed to trigger the gene silencing effect, it would be very advantageous to overcome the daunting challenge of high HIV mutations, which frequently generate new drug resistance to current small molecule drugs. Northern blot analysis of Figure 7A demonstrated the *ex-vivo* gene silencing effect of anti-HIV miRNA transfections (n = 3 for each set) on HIV-1 replication in $CD4^+$ T lymphocytes from both acute and chronic phase AIDS patients. In the acute phase (≤ one month), the 50 nM miRNA vector transfection degraded an average of 99.8% viral RNA genome (lane 4), whereas the same treatment knocked down only an average of 71.4±12.8% viral genome replication in the chronic phase (about 2-year infection). Immunocytochemistry staining of HIV p24 marker protein confirmed the results of Northern blot analysis (Figure 7B). Sequencing analysis has revealed that at least two HIV-1b mutants in the acute phase and seven HIV-1b mutants in the chronic phase were found within the targeted HIV genome domain. It is likely that the higher genome complexity of HIV mutations in chronic infections is able to counteract the miRNA-mediated silencing efficacy. Transfection of 50 nM miRNA* vector homologous to the HIV-1 genome failed to induce any RNAi effect on viral genome, indicating the specificity of the miRNA effect (lanes 5). Expression of cellular house-keeping gene, β-actin, was at a normal level and showed no interferon-induced non-specific RNA degradation. These results suggest that the designed anti-HIV SpRNAi-RGFP vector is highly specific and efficient in suppressing HIV-1 replication in the early infections. In conjunction with an intermittent interleukin-2 therapy [55], we may stimulate the growth of non-infected $CD4^+$ T lymphocytes to eliminate the HIV-infected cells.

The consistent evidence of miRNA-induced gene silencing effects in zebrafishes, chicken embryos, mouse skins and human diseases demonstrates the preservation of an ancient intron-mediated gene regulation system in eukaryotes. In these animal models, the intron-derived miRNAs determine the activation of RNAi-like gene silencing pathways. We herein provide the first time evidence for the biogenesis and function of intronic miRNAs *in vivo*. Given that natural evolution gives raise to more complexity and more variety of introns in higher animal and plant species for coordinating their vast gene expression volumes and interactions, dysregulation of these miRNAs due to intronic expansion or deletion will likely to cause genetic diseases, such as myotonic dystrophy and fragile X mental retardation. Thus, gene expression produces not only gene transcripts for its own protein synthesis but also intronic miRNAs, capable of interfering with other gene expressions. Based on this concept, the expression of a gene results in gain-of-function of the gene and also loss-of-function of some other genes, which contain complementarity to the mature intronic miRNAs. An array of genes can swiftly and accurately coordinate their expression patterns with each other through the mediation of their intronic miRNAs, bypassing the time-consuming translation processes under quickly changing environments. Conceivably, intron-mediated gene regulation may be as important as the mechanisms by which transcription factors regulate the gene expression. It is likely that intronic miRNA is able to trigger cell transitions quickly in response to external stimuli without the tedious protein synthesis. Undesired gene products are reduced by both transcriptional inhibition and/or translational suppression via miRNA regulation. This could enable a rapid switch to a new gene expression pattern without the need to produce various transcription factors. This regulatory property of miRNAs may serve as one of the most ancient gene modulation systems before the emergence of proteins. According to the variety of microRNAs and the complexity of genomic introns, a thorough investigation of miRNA variants in the human genome will markedly improve the understanding of genetic diseases and also the design of miRNA-based drugs. Learning how to exploit such a novel gene regulation system in the future therapy will be a forthcoming challenge.

References

[1] Clement, J.Q., Qian, L., Kaplinsky, N. and Wilkinson, M.F. (1999). The stability and fate of a spliced intron from vertebrate cells. *RNA 5*, 206-220.

[2] Lin, S.L., Chuong, C.M. and Ying, S.Y. (2001). A Novel mRNA-cDNA interference phenomenon for silencing bcl-2 expression in human LNCaP cells. *Biochem. Biophys. Res. Commun.* 281, 639-644.

[3] Lin, S.L., Chang, D., Wu, D.Y. and Ying, S.Y. (2003). A novel RNA splicing-mediated gene silencing mechanism potential for genome evolution. *Biochem Biophys Res Commun*. 310, 754-760.

[4] Ying, S.Y. and Lin, S.L. (2004). Intron-derived microRNAs–fine tuning of gene functions. *Gene* 342, 25-28.

[5] Ambros, V., Lee, R.C., Lavanway, A., Williams, P.T. and Jewell, D. (2003). MicroRNAs and other tiny endogenous RNAs in *C. elegans*. *Curr Biol*. 13, 807-818.

[6] Rodriguez, A., Griffiths-Jones, S., Ashurst, J.L. and Bradley, A. (2004). Identification of mammalian microRNA host genes and transcription units. *Genome Res*. 14, 1902-1910.

[7] Parrish, S., Fleenor, J., Xu, S., Mello, C. and Fire, A. (2000). Functional anatomy of a dsRNA trigger: differential requirement for the two trigger strands in RNA interference. *Mol. Cell 6*, 1077-1087.

[8] Holen, T., Amarzguioui, M., Wiiger, M.T., Babaie, E. and Prydz, H. (2002). Positional effects of short interfering RNAs targeting the human coagulation trigger Tissue Factor. *Nucleic Acids Res*. 30, 1757-1766.

[9] Hutvagner, G. and Zamore, P.D. (2002). A microRNA in a multiple-turnover RNAi enzyme complex. *Science* 297, 2056-2060.

[10] Zeng, Y., Yi, R. and Cullen, B.R. (2003). MicroRNAs and small interfering RNAs can inhibit mRNA expression by similar mechanisms. *Proc. Natl. Acad. Sci. USA* 100, 9779-9784.

[11] Hall, I.M., Shankaranarayana, G.D., Noma, K., Ayoub, N., Cohen, A. and Grewal, S.I. (2002). Establishment and maintenance of a heterochromatin domain. *Science* 297, 2232-2237.

[12] Llave, C., Xie, Z., Kasschau, K.D. and Carrington, J.C. (2002). Cleavage of Scarecrow-like mRNA targets directed by a class of Arabidopsis miRNA. Science 297, 2053-2056.

[13] Rhoades, M.W., Reinhart, B.J., Lim, L.P., Burge, C.B., Bartel, B. and Bartel, D.P. (2002). Prediction of plant microRNA targets. *Cell* 110, 513-520.

[14] Lee, R.C., Feibaum, R.L. and Ambros, V. (1993). The C. elegans heterochromic gene *lin-4* encodes small RNAs with antisense complementarity to *lin-14*. *Cell* 75, 843-854.
[15] Reinhart, B.J., Slack, F.J., Basson, M., Pasquinelli, A.E., Bettinger, J.C., Rougvie, A.E., Horvitz, H.R. and Ruvkun, G. (2000). The 21-nucleotide let-7 RNA regulates developmental timing in *Caenorhabditis elegans*. *Nature* 403, 901-906.
[16] Lau, N.C., Lim, L.P., Weinstein, E.G. and Bartel, D.P. (2001). An abundant class of tiny RNAs with probable regulatory roles in *Caenorhabditis* elegans. *Science* 294, 858-862.
[17] Brennecke, J., Hipfner, D.R., Stark, A., Russell, R.B. and Cohen, S.M.. (2003). Bantam encodes a developmentally regulated microRNA that controls cell proliferation and regulates the proapoptotic gene hid in *Drosophila*. Cell 113, 25-36.
[18] Xu, P., Vernooy, S.Y., Guo, M. and Hay, B.A. (2003). The Drosophila microRNA Mir-14 suppresses cell death and is required for normal fat metabolism. *Curr Biol.* 13, 790-795.
[19] Lagos-Quintana, M., Rauhut, R., Meyer, J., Borkhardt, A. and Tuschl, T. (2003). New microRNAs from mouse and human. *RNA 9*, 175-179.
[20] Mourelatos, Z., Dostie, J., Paushkin, S., Sharma, A., Charroux, B., Abel, L., Rappsilber, J., Mann, M. and Dreyfuss, G. (2002). miRNPs: a novel class of ribonucleoproteins containing numerous microRNAs. *Genes Dev.* 16, 720-728.
[21] Zeng, Y., Wagner, E.J. and Cullen, B.R. (2002). Both natural and designed micro RNAs can inhibit the expression of cognate mRNAs when expressed in human cells. *Mol Cell 9*, 1327-1333.
[22] Lin, S.L., Chuong, C.M. and Ying, S.Y. (2001). D-RNAi (messenger RNA-antisense DNA interference) as a novel defense system against cancer and viral infections. *Curr. Cancer Drug Targets 1*, 241-247.
[23] Carthew, R.W. (2001). Gene silencing by double-stranded RNA. *Curr Opin Cell Biol.* 13, 244-248.
[24] Lee, Y., Kim, M., Han, J., Yeom, K.H., Lee, S., Baek, S.H. and Kim, V.N. (2004). MicroRNA genes are transcribed by RNA polymerase II. *EMBO J.* 23, 4051-4060.
[25] Lee, Y., Ahn, C., Han, J., Choi, H., Kim, J., Yim, J., Lee, J., Provost, P., Radmark, O., Kim, S. and Kim, V.N. (2003). The nuclear RNase III Drosha initiates microRNA processing. *Nature* 425, 415-419.
[26] Lund, E., Guttinger, S., Calado, A., Dahlberg, J.E. and Kutay, U. (2004). Nuclear export of microRNA precursors. *Science* 303, 95-98.
[27] Yi, R., Qin, Y., Macara, I.G. and Cullen, B.R. (2003). Exportin-5 mediates the nuclear export of pre-microRNAs and short hairpin RNAs. *Genes Dev.* 17, 3011-3016.
[28] Schwarz, D.S., Hutvagner, G., Du, T., Xu, Z., Aronin, N. and Zamore, P.D. (2003). Asymmetry in the assembly of the RNAi enzyme complex. *Cell 115*, 199-208.
[29] Khvorova, A., Reynolds, A. and Jayasena, S.D. (2003). Functional siRNAs and miRNAs exhibit strand bias. *Cell* 115, 209-216.
[30] Lee, Y.S., Nakahara, K., Pham, J.W., Kim, K., He, Z., Sontheimer, E.J. and Carthew, R.W. (2004). Distinct roles for Drosophila Dicer-1 and Dicer-2 in the siRNA/miRNA silencing pathways. *Cell* 117, 69-81,

[31] Liquori, C.L., Ricker, K., Moseley, M.L., Jacobsen, J.F., Kress, W., Naylor, S.L., Day, J.W. and Ranum, L.P.W. (2001). Myotinic dystrophy type 2 caused by a CCTG expansion in intron 1 of ZNF9. *Science* 293, 864-867.

[32] Jin, P., Alisch, R.S. and Warren, S.T. (2004). RNA and microRNAs in fragile X mental retardation. *Nat Cell Biol.* 6, 1048-1053.

[33] Eberhart, D.E., Malter, H.E., Feng, Y. and Warren, S.T. (1996). The fragile X mental retardation protein is a ribonucleoprotein containing both nuclear localization and nuclear export signals. *Hum Mol Genet.* 5, 1083-1091.

[34] Tuschl, T. and Borkhardt, A., (2002). Small interfering RNAs: a revolutionary tool for the analysis of gene function and gene therapy. *Mol Interv.* 2, 158-167.

[35] Miyagishi, M. and Taira, K., (2002). U6 promoter-driven siRNAs with four uridine 3' overhangs efficiently suppress targeted gene expression in mammalian cells. *Nat Biotechnol.* 20, 497-500.

[36] Lee, N.S., Dohjima, T., Bauer, G., Li H., Li, M.J., Ehsani, A., Salvaterra, P. and Rossi, J. (2002). Expression of small interfering RNAs targeted against HIV-1 rev transcripts in human cells. *Nat Biotechnol.* 20, 500-505.

[37] Paul, C.P., Good, P.D., Winer, I. and Engelke, D.R. (2002). Effective expression of small interfering RNA in human cells. *Nat Biotechnol.* 20, 505-508.

[38] Xia, H., Mao, Q., Paulson, H.L. and Davidson, B.L. (2002). siRNA-mediated gene silencing in vitro and in vivo. *Nat Biotechnol.* 20, 1006-1010.

[39] McCaffrey, A.P., Meuse, L., Pham, T.T., Conklin, D.S., Hannon, G.J. and Kay, M.A. (2002). RNA interference in adult mice. *Nature* 418, 38-39.

[40] Gunnery, S., Ma, Y. and Mathews, M.B. (1999). Termination sequence requirements vary among genes transcribed by RNA polymerase III. *J Mol Biol.* 286, 745-757.

[41] Schramm, L. and Hernandez, N. (2002). Recruitment of RNA polymerase III to its target promoters. *Genes Dev.* 16, 2593-2620.

[42] Sledz, C.A., Holko, M., de Veer, M.J., Silverman, R.H. and Williams, B.R. (2003). Activation of the interferon system by short-interfering RNAs. *Nat Cell Biol.* 5, 834-839.

[43] Lin, S.L. and Ying, S.Y. (2004). Combinational therapy for HIV-1 eradication and vaccination. *Intrn'l J. Oncol.*. 24, 81-88.

[44] Stark, G.R., Kerr, I.M., Williams, B.R, Silverman, R.H. and Schreiber, R.D. (1998). How cells respond to interferons. *Annu. Rev. Biochem.* 67, 227-264.

[45] Lin, S.L. and Ying, S.Y. (2004). New drug design for gene therapy – Taking Advantage of Introns. *Lett Drug Design and Discovery* 1, 256-262.

[46] Lin, S.L. and Ying, S.Y. Novel RNAi therapy – Intron-derived microRNA drugs. *Drug Design Reviews 1*, 247-255, 2004.

[47] Nott, A., Meislin, S.H. and Moore, M.J. (2003). A quantitative analysis of intron effects on mammalian gene expression. *RNA* 9, 607-617.

[48] Zhang, G., Taneja, K.L., Singer, R.H. and Green, M.R. (1994). Localization of pre-mRNA splicing in mammalian nuclei. *Nature* 372, 809-812.

[49] Ghosh, S. and Garcia-Blanco, M.A. (2000). Coupled in vitro synthesis and splicing of RNA polymerase II transcripts. *RNA* 6, 1325-1334.

[50] Butz, S. and Larue, L. (1995). Expression of catenins during mouse embryonic development and in adult tissues. *Cell Adhes Commun.* 3, 337-352.

[51] Filipovska, J. and Konarska, M.M. (2000). Specific HDV RNA-templated transcription by pol II in vitro. *RNA* 6, 41-54.

[52] Modahl, L.E., Macnaughton, T.B., Zhu, N., Johnson, D.L. and Lai, M.M. (2000). RNA-Dependent replication and transcription of hepatitis delta virus RNA involve distinct cellular RNA polymerases. *Mol Cell Biol.* 20, 6030-6039.

[53] Abzhanov, A., Protas, M., Grant, B.R., Grant, P.R. and Tabin, C.J. (2004). Bmp4 and morphological variation of beaks in Darwin's finches. *Science* 305, 1462-1465.

[54] Wu, P., Jiang, T.X., Suksaweang, S., Widelitz, R.B. and Chuong, C.M. (2004). Molecular shaping of the beak. *Science* 305, 1465-1466.

[55] Kovacs, J.A., Vogel, S., Albert, J.M., Falloon, J., Davey, R.T.Jr., Walker, R.E., Polis, M.A., Spooner, K., Metcalf, J.A., Baseler, M., Fyfe, G. and Lane, H.C. (1996). Controlled trial of interleukin-2 infusions in patients infected with the human immunodeficiency virus. *N. Engl. J. Med.* 335, 1350-1356.

In: Gene Silencing: New Research
Editor: Grace W. Redberry, pp. 165-191
ISBN 1-59454-832-3

Chapter VI

DNA Methylation and Epigenetic Gene Silencing in Breast Cancer

Qifeng Yang,[1,*] Guohong Hu,[2] Zhigang Yu,[3] Jingzhong Sun,[4] Goro Yoshimura[5] and Kennichi Kakudo[6]

[1] The Cancer Institute of New Jersey, New Brunswick, NJ;
[2] Princeton University, Princeton, NJ;
[1,3,4] Qilu Hospital of Shandong University, Ji'nan, P.R. China;
[1,5] Affiliated Kihoku Hospital of Wakayama Medical University, Japan;
[1,6] Wakayama Medical University, Japan

Abstract

Of all epigenetic modifications, hypermethylation is commonly considered as a critical causal event in silencing important classes of genes in cancer. In breast cancer, a series of tumor suppressor or related genes are frequently hypermethylated and which is associated with silencing of these genes. We now understand that DNA methylation mechanisms might include a number of DNA methyltransferases, proteins that direct DNA methyltransferases to specific promoters, chromatin modifying enzymes as well as demethylases. There are also data to suggest that pharmacological down regulation of some members of the DNA methylation machinery could inhibit cancer *in vitro, in vivo* and in clinical trials. The use of the methylation status of certain genes as a marker of cancer or cancerous conditions provides an additional weapon in early detection and prognosis of breast cancers. We describe the techniques used to decipher the complement of methylation-inactivated genes, discuss the epigenetic regulation of certain critical tumor suppressor and growth regulatory genes implicated in breast cancer and its relevance to breast cancer development, progression and therapy.

* Address correspondence to Qifeng Yang, M.D. Ph.D., The Cancer Institute of New Jersey, 195 Little Albany St., New Brunswick, NJ 08903-2681, USA, e-mail: yang-qf@mail.wakayama-med.ac.jp

Introduction

Breast cancer is the most common cancer among women, besides skin cancer. It is the second leading cause of cancer death in women, after lung cancer. About 211,240 women in the United States will developp invasive breast cancer in 2005. Although breast cancer mortality has begun to decline because of improved screening techniques, About 40,410 women will die from the disease this year. The transformation of normal cells into cancer cells involves multiple etiologic pathways and, at the very least, is a genetic, epigenetic and cytogenetic process. The genetic route to tumorigenesis has been studied extensively, and epigenetics is now coming of age. The genetic and epigenetic alterations that initiate and drive tumorigenesis are promising targets for early detection because they may precede clinically obvious cancers, can be detected at sensitive levels, may be specific to tumor cells, and can potentially provide information about the prognosis and treatment of the disease [1, 2].

The term "epigenetics" defines all meiotically and mitotically heritable changes in gene expression that are not coded in the DNA sequence itself. Three systems, including DNA methylation, histone modification and RNA-associated silencing, are used to initiate and sustain epigenetic silencing. DNA methylation is the only commonly occurring modification of human DNA [3]. Histone modification includes acetylation and methylation of conserved lysine residues on the amino-terminal tail domains. The acetylation of histones marks active, transcriptionally competent regions, whereas hypoacetylated histones are found in transcriptionally inactive euchromatic or heterochromatic regions. Histone methylaiton can be a marker for both active and inactive regions of chromatin. Antisense transcripts, noncoding RNAs (such as Xist) or RNA interference (RNAi), can also lead to mitotically heritable transcriptional silencing by the formation of heterochromatin. The major form of epigenetic information in mammalian cells is DNA methylation, or the covalent addition of a methyl group to the 5-position of cytosine predominantly within the CpG islands. DNA methylation has been shown by several lines of evidence to play a role in gene activity, cell differentiation, tumorogenesis, X-chromosome inactivation, genomic imprinting and other major biological processes. DNA hypermethylation has been used as a biomarker for breast cancer risk, classification and prognosis, and it is also a new target for breast cancer treatment.

Methyltransferase, Methylation and Gene Transcription

DNMTs and Methylation

The major form of epigenetic information in mammalian cells is DNA methylation, or the covalent addition of a methyl group to the 5-position of cytosine predominantly within the CpG islands. CpG islands are regions of more than 500 base pairs in size and with a GC content greater than 55% [4], CpG islands have been conversed during evolution because they are normally kept free of methylation. CpG islands are located within the promoter regions of about 40% of mammalian genes and, when methylated, cause stable heritable transcriptional silencing. Aberrant *de novo* methylation of CpG islands is a hallmark of human cancers and is found early during carcinogenesis.

DNA methylation is accomplished through the activity of specific enzymes, the DNA methyltransferases (DNMT's). The DNMTs known to date are DNMT1, DNMT1b, DNMT1o, DNMT1p, DNMT2, DNMT3A, DNMT3B with its isoforms, and DNMT3L[5]. Methylation can be *de novo* (when CpG dinucleotides on both DNA stands are unmethylated) or maintenance (when CpG dinucleotides on one strand are methylated). DNMT1 has *de novo* as well as maintenance methyltransferase activity, and DNMT3A and DNMT3B are powerful *de novo* methyltransferases [6]. DNMT2 has been cloned and characterized but catalytic activity of this enzyme has yet to be demonstrated *in vitro* or *in vivo* [7], but its catalytic domain lacks DNMT activity in the human.

DNMT1, the first methyltransferase, was discovered by Bestor et al [8]. The DNMT1 gene is located at human chromosome 19p13.2 and encodes a 200 kDa protein whose methyltransferase catalytic domain is located at the C-terminus of the protein. The generation of DNMT1-knockout mice has revealed that DNMT1 is required for proper embryonic development, imprinting, and X-inactivation. Mice deficient for DNMT1 show arrested development prior to the 8 somite stage and a nearly 70% reduction in their genomic 5-methylcytosine content. A research group at Johns Hopkins University [9] reported that DNMT1 expression is dysregulated in breast cancer. The elevated protein levels are not a result of increased mRNA levels, but rather an increase in protein half-life. They found that DNMT1 protein levels were elevated in breast cancer tissues and in MCF-7 breast cancer cells relative to normal Human Mammary Epithelial Cells (HMECs) without a concomitant increase in DNMT1 mRNA or proliferative fraction. While DNMT1 mRNA levels were properly S-phase regulated in both cell types, DNMT1 protein levels did not follow S-phase fraction in MCF-7 cells. Rather, an increase in DNMT1 protein stability was found for MCF-7 cells relative to HMECs, and a destruction domain was mapped to the N-terminal 120 amino acids of DNMT1 which was required for its proper ubiquitination and degradation in HMECs. Furthermore, overexpression of DNMT1 with this deleted destruction domain in HMECs resulted in significantly increased genomic 5-methylcytosine levels relative to overexpression of the full-length protein. The regulation of DNMT1 destruction via this domain may be dysfunctional in cancer cells leading to subsequent cytosine hypermethylation in the genome. Okano *et al* first described the characterization of the DNMT3 family of methyltransferases [10]. DNMT3 family members share some structural similarities with DNMT1 with a catalytic C-terminal region responsible for DNA methylation and a N-terminal region regulatory domain. They methylate CpG dinucleotides of unmethylated and hemimethylated DNA *in vitro*. The two genes are expressed at high levels in ES cells and relatively low levels in adult somatic tissues. Human DNMT3A has been mapped to chromosome 2p23 whereas DNMT3B maps to chromosome 20q11.2 [11]. Human DNMT3A is ubiquitously expressed but DNMT3B is expressed at low levels except in testis, thyroid and bone marrow. Overexpression of both DNMT3B and DNMT3A appears to characterize multiple types of human tumors [12]. A specific function for Dnmt3b, revealed by the mouse knockout experiments, is maintenance of DNA methylation of the minor satellite repeats adjacent to entromeres [13, 14]. The specific function for DNMT3A was not detectable with the knockout model but studies using the various DNMT-knockout ES cells in addition to transgenic Drosophila melanogaster expressing DNMT3A revealed that this enzyme may be specialized to methylate non-CpG sequences like CpA, and CpT although the function of non-CpG methylation in ES cells is unknown [15]. The catalytic activity of DNMT3A and 3B are much lower than that of DNMT1 [10]. DNMT1 is mainly responsible for maintenance of DNA methylation, whereas DNMT3A and DNMT3B have been shown to methylate hemimethylated and unmethylated DNA with equal efficiencies [13]. Overexpression of DNMT1 as well as the DNMT3 mRNAs has been reported in human tumors [11, 16], but this phenomenon is probably only partially responsible for the observed methylation changes.

Null mutations of the mammalian DNMTs are lethal at different stages, either prenatally and postnatally. ES cells homozygous for DNMT1 mutation are viable, with no growth or retain the 30% of genomic methylation. Some studies on cells with total or partial loss of DNMT1 function revealed abnormalities in transcriptional silencing of specific imprinted genes and the Xist allele on the chromosome [17]. Mice lacking DNMT3A die postnatally, whereas those lacking DNMT3B develop normally until E9.5, then die prior to term. Double mutants already show abnormality at E8.5, then die prior to E11.5. Specific DNMT isoforms, such the DNMT1o, oocyte-specific seem to play specific roles in the physiology of genomic methylation [18].

Methylation and Gene Transcription

The link between methylation and gene expression was revealed more than two decades ago when a mouse-human somatic cell hybrid clone was isolated [19], however, evidence directly connecting the two events has only recently been obtained [20].

The first mechanism suggested that the suppression of transcription by DNA methylation proposed direct inhibition of transcription factor binding. The active recruitment of methyl-CpG binding activities appears to be the dominant mechanism of methylation-dependent repression. MeCP1 and MeCP2 were the first two methyl-CpG binding activities to be described [21]. MeCP1 was originally identified as a large multi-protein complex, MeCP2 is a single polypeptide with an affinity for a single methylated CpG. MeCP2 germline mutations are responsible for the mental retardation disease Rett syndrome, and the absence of a functional MeCP2 in these patients causes specific dysregulation of the gene expression. Additional proteins harbouring the methyl-CpG binding domain (MBD), namely MBD1, MBD2, MBD3, and MBD4 have also been identified [22]. The first molecular evidence concerning the role of methyl-CpG binding proteins in repressing aberrantly methylated genes in cancer arose from the demonstration that methylation of a construct containing part of the 5' BRCA1 CpG island inhibited gene expression in the presence of MeCP2 [23]. Recently, the presence of MBD proteins has been demonstrated to be a major hallmark of tumor suppressor genes hypermethylated and silenced in cancer [24].

Methylation might also influence transcription indirectly through chromatin condensation, which is the packing of DNA into high-order three-dimensional protein-nucleic acid structures, often associated with gene inactivity. The connection between DNA methylation and transcriptional repression by the modification of chromatin was first revealed in 1997 [20, 25]. Chromatin research has resulted in the description of a variety of histone modifications and the specificity of certain modifications under particular physiological conditions. Most histone modifications occur in their protruding N-terminal tails. It is hypothesized that histone modifications act sequentially or in combination to form a code that may be read by nuclear factors [26]. There are several modifications that are compatible with gene silencing. In general, histone deacetylation leads to gene silencing, whilst histone acetylation leads to gene activation. Although cytosine methylation can directly inhibit the binding of certain transcription factors, methylated DNA is recognized by different methyl-DNA binding proteins, such as MeCP2 and MBD2. It is widely accepted that methylated DNA binding proteins such as MeCP2 target methylated DNA. MeCP2 recruits co-repressor complexes and histone deacetylases and histone methyltransferases to the methylated gene precipitating an inactive chromatin structure [20, 25, 27]. Another important issue is that histone acetyl transferases and deacetylaces remodel chromatin by increasing or decreasing the net charge of the histones themselves, thus altering their affinity for negatively charged DNA. This process could alter gene expression by making DNA either more or less accessible to transcriptional activators or repressors.

Epigenetic modification may be found on the DNA primary sequence (e.g., cytosine methylation) or on the associated histones (lysine or arginine acetylation, serine phosphorylation), which determine the balance between chromatin activation or inactivation. Although it is now accepted that DNA methylation causes histone deacetylation, it is still not clear if histone acetylation can cause local DNA demethylation. DNA methylation could drive heterochromatin formation through other pathways independent of histone deacetylation. Recently, it was discovered that all three DNA methyltransferases interact directly with histone deacetylases HDAC1 and HDAC2 [28], inducing herochromatin formation and repressing transcription, bypassing cytosine methylation [29]. On the other hand, genes without a CpG island in their promoter cannot be efficiently silenced by methylation, but their transcription correlates with levels of histone acetylation in their promoters.

Short interfering RNAs (siRNAs) may play a critical role in gene silencing in the nucleus. SiRNAs are small duplex RNA molecules with a 2-bp 3' overhang, which have long been known to plant geneticists as mediators of post-transcriptional gene silencing (PTGS). PTGS, also known as 'RNA interference (RNAi)', consists of degradation of mRNAs homologous to the siRNAs [30]. SiRNA are able to induce heterochromatinization of homologous DNA stretches by directing DNA methylation and/or other chromatin modifications. One such modification is the methylation of histone H3 K9 residue. Recently, Kawasaki *et al.* [31] indicated that RNAi can cause transcriptional gene-silencing by RNA-directed promoter methylation in human breast cancer and in non-transformed breast epithelial cells. They presented evidence that siRNAs targeting CpG islands in the E-cadherin promoter increased methylation in MCF7 breast cancer cells, and they showed this was an RNAi sequence-specific event, and that a similar process could be induced in normal breast epithelial cells. Notably, histones can also be methylated by these RNAi sequences. The concentration of E-cadherin mRNA is decreased by using several siRNAs, and this siRNA-induced methylation was mediated by DNA methylases 1 and 3B. Although much work remains to be done in order to characterize naturally occurring short RNAs, antisense transcripts, and proteins interacting with short RNAs, it is clear that RNA plays a critical role in directing the epigenetic status of the genome.

Aberrant Methylation, Chromosome Instability and Mismatch Repair

Two kinds of changes in the DNA methylation pattern occur in breast cancer, regional hypermethylation of certain genes [32] and global hypomethylation [33, 34]. The exact nature of the defect in the cellular methylation machinery in tumor cells remains unknown. The global hypomethylation in cancer can not result from lack of DNMT activity or lack of methyl donors since many genes are hypermethylated and there is no evidence that DNMT levels are decreased in cancer, on the contrary several reports show elevated and deregulated levels of DNMT1 in cancer [35, 36]. Another hypothesis is that there is a global increase in demethylation activity in cancer cells, which drives global hypomethylation [37]. It is important to stress that tumors exhibit a global defect in the DNA methylation system with hypo- and hypermethylation events occurring in the same cells. DNA methylation changes may ultimately lead to the genetic instability characteristic of cancer in several ways.

First, 5 methylcytosine (5meC) serves as site of transition mutation by the hydrolytic deamination of 5meC to thymine. This phenomenon was used to explain the high incidence of CpG to TpG transition mutations observed in the p53 tumor suppressor gene. Such mutation frequently occurs in the well-known p53 tumor suppressor gene [38] and has also been reported in several other important genes like Rb, and c-H-ras-1 [39]. The relative importance of spontaneous deamination of 5-methylcytosine and carcinogen adduct formation at CpG sites in producing mutations remains to be clarified. Secondly, hypermethylation of certain critical genes in cancer by promoter methylation may predispose to genetic instability. Methylation of the mismatch repair gene (MLH1) is thought to be functionally linked with microsatellite instability (MSI) [40]. Mice knockouts of MLH1 show increased spontaneous mutations and cancer risk [41]. A close correlation between mismatch repair, genetic instability and methylation capacity has been reported in colon cancer cell models [42] [43]. In addition, a p53-inducible gene, 14-3-3σ is methylated in many cancers including breast carcinoma. Its inactivation may also facilitate the accumulation of genetic damages [44]. Thirdly, hypomethylation of repetitive sequences may also predispose cells to chromosomal defects and rearrangements that result in genetic instability [45]. The loss of methylation in tumor cells may predispose them to genomic instability *via* the transcriptional activation and movement of parasitic elements, these elements also pose a significant threat to the genome as mediators of homologous recombination. The most direct evidence that DNA methylation suppresses homologous recombination has come from the work in Ascobolus, but several studies support a similar role in mammalian cells. For example, DNMT1 knockout ES cells exhibit a 10-fold increased mutation rate involving gene rearrangements [46], and individuals with ICF syndrome or cultured cells treated with demethylation agent show increased numbers of chromosomal translocations [47].

Epigenetic change can predispose critical genetic mutations often found in tumorigenesis [48]. Methylation-mediated silencing occurs before the genetic change that underlies the MSI phenotype in endometrial cancers [49]. In breast cancer, it is postulated that methylation-associated inactivation of GSTP1 can result in adenine or guanine mutation by estrogen metabolites-DNA adduct formation and lead to genetic instability [50]. Hypermethylation and hypomethylation events can occur concomitantly. Rearrangements in the centromere of chromosomes 1 and 16 are observed in Wilms tumors and are thought to contribute to disease progression [51]. In breast carcinoma, satellite 2 DNA exhibits decreased methylation whereas the E-cadherin promoter is hypermethylated [51]. It is assumed that DNA methylation may suppress homologous recombination *via* masking of the recombination initiation site, maintenance of a highly condensed chromatin structure through the recruitment of DNA organizing proteins, destabilization of the recombination intermediate, or interference with the assembly of the recombination machinery. Loss of DNA methylation from repetitive elements may occur early in tumor progression and may predispose cells to genome rearrangements *via* mitotic recombination which could inactivate critical growth regulatory genes or, aberrantly regulate tumor suppressor genes or oncogenes.

Methods for Studying DNA Methylation

Methylation can be monitored either by methods that require methylation-sensitive restriction enzymes, or by methods that rely on chemical modification of the DNA with sodium bisulfite. Restriction enzyme based methods are based on the inability of methylation sensitive restriction enzymes to cleave methylated cytosines in their recognition site (Figure 1). After cleavage, analysis of methylation patterns has traditionally been conducted by utilizing southern hybridization techniques or PCR and is based on the length of the digested DNA fragment. The inability to digest methylated sequences results in longer fragments, indicating a methylated CpG dinucleotide. Restriction enzyme based methods are simple and rapid, however, it is limited to specific restriction sites and a substantial amount of initial DNA is required. This technique is useful for genome-wide methylation analyses and marker discovery. False positive results may result from incomplete digestion by the enzymes. Chemical modification of the DNA with sodium bisulfite for methylation analysis was discovered by Frommer et al. in 1992 [52](Figure 2). Sodium bisulfite reacts with cytosines in a methylation-specific manner: unmethylated cytosines are converted to uracils and methylated cytosines are not (Figure 3). The difference in methylation status marked by bisulfite reactivity can be accurately determined by PCR, sequencing, array hybridization, etc. Sodium bisulfite based detection of DNA methylation can circumvent the problems associated with restriction enzyme analysis and is now used extensively.

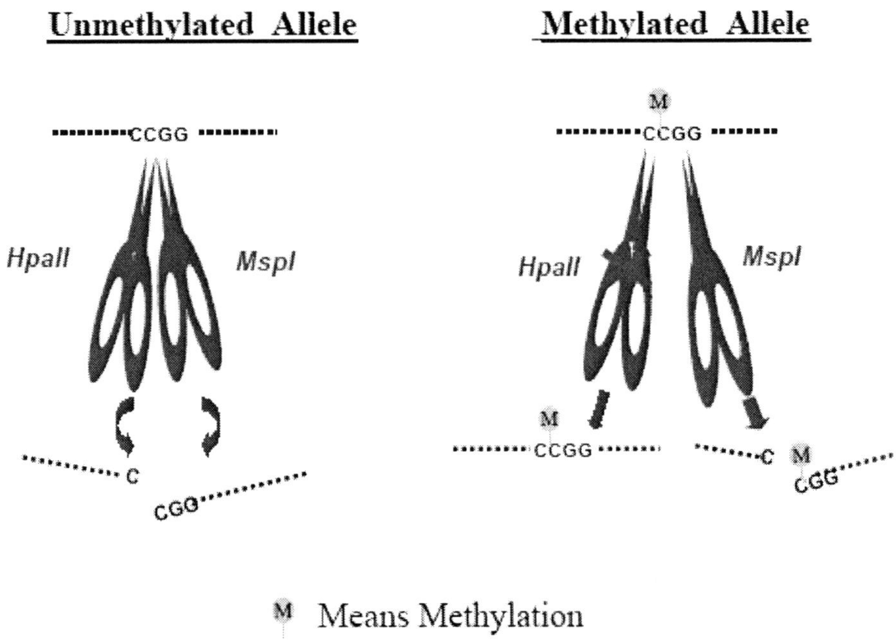

Figure 1. The cleavage of DNA by restriction endonucleases HpaII and MSPI. HapII is prevented by the presence of a 5-methyl group at the internal C residue of its recognition sequence CCGG. MspI cleaves DNA irrespective of the presence of a methyl group at this position.

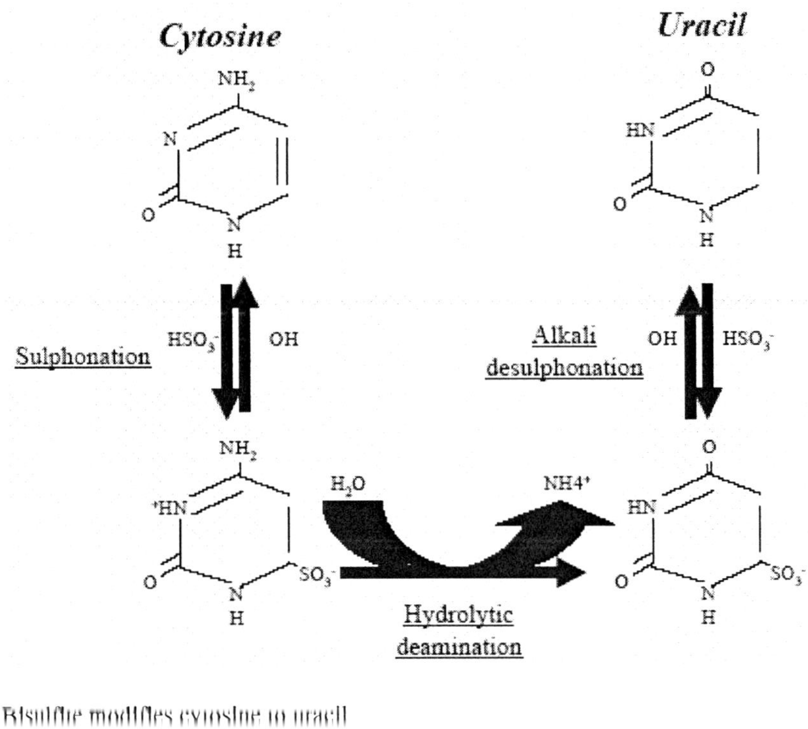

Figure 2. Bisulfite modifies cytosine to uracil

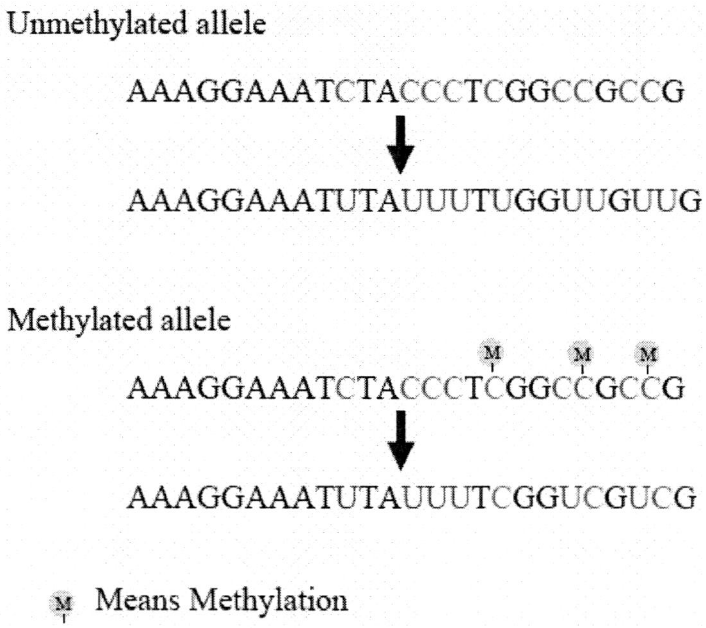

Figure 3. Sequence change with bisulfite treatment. In unmethylated allele, unmethylated cytosines are converted to uracils. In methylated allele, only unmethylated cytosines are converted to uracils.

Methylation-specific PCR (MSP) was introduced by Herman *et al.* [53]. MSP is based on the use of two distinct methylation specific primer sets for the sequence of interest. The primers must be specific for sodium bisulfite converted DNA and therefore have to contain a substantial number of both non-CpG and CpG cytosines in the original template. Both unmethylated (U) and methylated (M) primers are commonly designed for analysis. The U primer will only amplify sodium bisulfite converted DNA in an unmethylated condition, while the M primer is specific for sodium bisulfite converted methylated DNA. Software tools are available to support MSP primer design [54, 55]. MSP provides a sensitive and cost-effective test and makes it possible to analyze clinical samples with a limited amount of available template DNA. To increase the sensitivity of MSP for detection of methylated DNA in a background of unmethylated DNA, Palmisano *et al.* [56] developed a nested, two-round MSP. The first round primer is designed in regions flanking the MSP primers and harbouring a substantial number of cytoines but no CpG dinucleotides. In this way, the first round PCR should amplify both methylated and unmethylated alleles. The first round PCR product is subjected to second round PCR reactions using U and M specific primers as described above. Using nested MSP, one methylated allele in >50000 unmethylated alleles is detectable. False positives for MSP analysis may be caused by the lack of the single-stranded conformation, renaturation and incomplete sodium bisulfite conversion [57-59]. Commercially available kits can improve the ease and robustness of bisulfite treatment (e.g., EZ DNA methylation kit by *Biocompare Inc.* [South San Francisco, CA] and *Zymco Research Corp.* [Orange, CA] CpGenome modification kit by *Chemicon International* [Temecula, CA] and *Intergen Co.* [Burlington, MA]. Nevertheless, when using bisulfite treatment, reaction conditions must be well controlled and controls must ensure that the treatment is complete but has not resulted in excessive degradation of DNA. Commercial and noncommercial software can help in the design of PCR primers for bisulfite-treated genomic DNA (e.g., CpGWARE primer design software by Chemicon International and Methprimer by the University of California, San Francisco Department of Urology, [available at www.urogene.org/methprimer/]). Finally, the amplification of bisulfite-treated DNA may result in preferential amplification of either the methylated or unmethylated alleles.

Quantitative MSP (Q-MSP) technology which takes advantage of the development of quantitative real-time PCR has been developed to exactly quantify the number of methylated alleles. A well-known assay is MethyLight [60]. This method relies on relative quantification against controls that are methylation insensitive, as well as a standard curve based on serial dilution of a control sample. A typical MethyLight reaction has two primers and a probe, each incorporating several CpGs so that the assay is specific. During the extension phase of the PCR, the 5' to 3' exonuclease activity of the TaqDNA Polymerase cleaves the reporter from the probe, releasing it from the quencher, resulting in an increase in fluorescence emission. Data are normalized to an internal reference gene, *e.g.* actin, which is designed in a region that does not contain any CpG nucleotides to allow an unbiased amplification. It should be noted that the quantitation provided is for the relative amounts of methylated DNA with a particular methylation pattern compared to some reference control gene. This quantitation may reflect actual changes in methylation density or differences in tumor DNA within a specific sample. To further increase the sensitivity of this technology, Fackler *et al.* [61] used a quantitative multiplex nested MSP (QM-MSP) strategy. External primers lacking CG nucleotides are used in the first round PCR reaction, therefore, both methylated and unmethylated alleles are amplified. In the second step, quantitative analysis of methylated and unmethylated DNA for each gene was performed separately with the primer pairs and probes. Q-MSP has been used for analyzing formalin-fixed, paraffin-embedded surgical samples even laser-assisted microdissected samples of breast tumors [62]. Other analytical methods include the use of methylation-specific restriction enzymes, methylation-specific single-base primer extension, arrays, or Sanger sequencing. The latter can be applied as direct genomic sequencing for semiquantitative analysis of multiple adjacent CpG sites, setting it apart from most methods, which generate data for an individual position per assay (i.e., primer). Pyrosequencing technology is a method for sequencing-by-synthesis in real time. It is based on an indirect bioluminometric assay of the pyrophosphate (PPi) that is released from each deoxynucleotide (dNTP) upon DNA-chain elongation. Pyrosequencing technology has a number of valuable attributes for methylation studies. This method provides internal controls to confirm the completeness of bisulfite treatment together with the ability to analyze several individual neighboring methylation sites. In addition, the surrounding sequence generated around the methylated areas provides control peaks that can be used to confirm the correct placement of the primer on the modified DNA template, which is now rich in adenine (A) and thymine (T).

HeavyMethyl assay [63] is another PCR-based technique for methylation detection. This kind of assay uses a blocker oligo to prevent priming of unmethylated DNA, therefore, a template with at least one methylated CpG in the blocking area and comethylation of the CpGs in the probe binding region is required. Recently, the genome-wide methylation profiling microarray technique has been able to combine efficiently either the fragmentation by a methyl-sensitive restriction endonuclease followed by size fractionation [64] or the MSP methodology described above with the high density array method [65]. Different microarray-based strategies that have been used for methylation profiling have distinct features and provide different types of information. Ednonuclease cleavage combined with either size fractionation or PCR amplification typically has a mapping resolution of a few kilobases, without providing specific information about any individual methylated base. Therefore, cleavage-based methods are suitable for surveying gene-sized targets. One approach uses ligation-mediated PCR to preferentially amplify strongly methylated genomic regions [66]. The amplification of nonmethylated regions is suppressed by the digestion of the template DNA before PCR with the restriction enzymes BstUI and HpaII, which are blocked by cytosine methylation. A related method [67] uses digestion with SmaI, which cleaves the unmethylated sequence 5'-CCCGGG-3', followed by digestion with XmaI. The cleavage products of XmaI have 5' protruding ends, whereas SmaI leaves blunt ends. Following adaptors specific for the XmaI ends are ligated, PCR is performed using primers that are complementary to these adaptors. The PCR product should contain only genomic regions with methylated 5'-CCCGGG-3' sequences, could be hybridized to microarrays. In contrast, the bisulfite-based strategy, which has a single base pair resolution, provides only information about specific residues that have been chosen in advance as being informative. Therefore, the bisulfite-based methods may be used when particular sites have been identified and oligonucleotide pairs optimized.

Hypermethylation as Breast Cancer Biomarkers

DNA hypermethylation is the most common and best characterized epigenetic abnormality in human malignancies, including breast cancer. By searching the Medline database using query "breast AND hypermethylation", 222 records appeared on March 8, 2005. These genes include classic and putative tumor-suppressor genes and genes involved in a number of cellular pathways such as hormonal responses, tumor-cell invasion and/or tumor architecture, cell cycle control, and DNA damage repair. For many of these genes, promoter hypermethylation is often the main mechanism underlying the functional loss in breast cancer. Aberrant silencing of these genes can contribute to cancer initiation, progression, and metastasis. Hypermethylation of these genes in breast cancer can correlate with histological type, pathologic grade, clinical stage and prognosis.

Hypermethylation as a Marker for Predicting Breast Cancer Risk

To date, several studies have shown that methylation-associated silencing inactivates certain tumor suppressors as effectively as mutations and is one of the cancer-predisposing hits described in Knudson's two hit hypothesis. For some genes, an aberrant decrease or absence of expression may not be a consequence of somatic mutations or allele loss at all, DNA hypermethylation may be the primary mechanism of gene inactivation. One possibility is that hypermethylation could begin in a monoallelic form with subsequent expansion to both alleles of the gene and the resultant complete abrogation of gene function. Methylation may also function synergistically with genetic mutations to promote tumourigenesis in a more indirect fashion. In our previous study, we demonstrated that biallelic inactivation of the FHIT gene could result from either epigenetic inactivation of both parental alleles, or from epigenetic modification of one allele and deletion of the remaining allele [68]. A model for the biallelic inactivation of tumor suppressor genes by aberrant DNA methylation alone or in combination with mutations or deletions is shown in Figure 4.

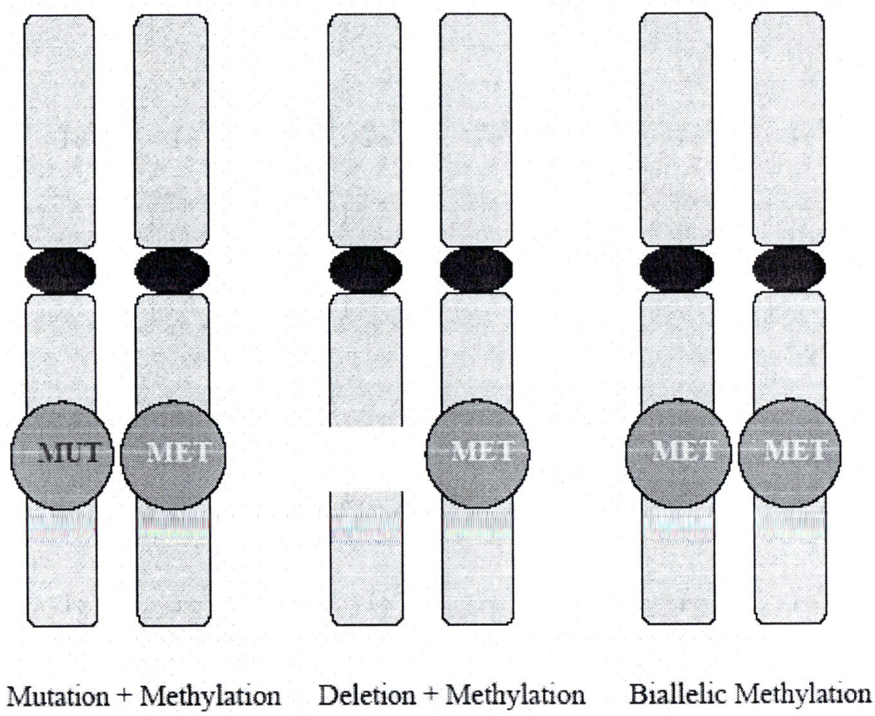

Figure 4. Schematic representation of tumor suppressor gene inactivation. Hypermethylation (METH) progresses in a monoallelic form and can expand to both alleles of the gene. Biallelic inactivation of tumor suppressor gene may be resulted from mutation (MUT) and hypermethylation, loss of heterozygosity (deletion) and hypermethylation, and biallelic hypermethylation.

Epidemiologic risk factors such as age, cancer family history, history of breast biopsies, and reproductive history have been combined in mathematical models that are useful for categorizing women as having a low, intermediate, or high risk for breast cancer. The tumor suppressor genes (TSGs) RASSF1A, APC, H-cadherin, RARβ2, and cyclin D2 are methylated more frequently in breast cancer than in adjacent benign tissue. These data suggest that methylation of certain TSGs may occur as a field change and may be associated with increased risk for malignant progression.

Lewis *et al.* [69] analyzed the methylation status of these five TSGs in benign and malignant breast cells to determine whether TSG methylation frequency in benign breast tissue is associated with epidemilogic markers of breast cancer risk. They found that cyclin D2 methylation seems to occur later in breast epithelial cell transformation and may be exploited in the future as a marker for the early detection of breast cancer, whereas RARβ2,

APC, and RASSF1A promoter methylation occurs frequently in benign breast epithelium and is associated with epidemiologic markers of increased breast cancer risk. More specifically, RARβ2 methylation is correlated with a personal history of breast cancer, whereas APC and RASSF1A methylation is associated with calculated breast cancer risk particularly in women with prior benign breast biopsies. This may suggest a field change in benign breast epithelium that could be exploited for tissue-based individualized breast cancer risk stratification.

Hypermethylation as a Marker for Cancer Detection

Detection of breast cancer at an early stage is the key to successful treatment and outcome. Despite the availability of mammography and prevalence of self-examination, there is still additional benefit to be gained from additional screening methodologies. Novel approaches to early detection of breast neoplasia therefore merit examination. The genetic and epigenetic alterations that initiate and drive tumorigenesis are promising targets for early detection because they may precede clinically obvious cancer, can be detected at sensitive levels, may be specific to tumor cells, and can potentially provide information about the prognosis and treatment of the disease.

Currently, fine needle aspiration (FNA) cytology is an integral part of the diagnostic evaluation of suspicious breast lesions. However, because morphological features of malignancy are more ambiguous in small tumors and premalignant lesions, this procedure is prone to challenge the cytopathologist, and renders a diagnosis that is highly dependent on the proficiency of the performer. Nipple aspiration is a noninvasive technique that provides a fluid sample containing both cells shed from the breast duct lining and proteins secreted and concentrated into the duct lumen. FNA washings from breast lesions and nipple aspiration are suitable clinical samples for DNA-based detection of breast cancer. Detection of hypermethylation offers several advantages. A major one is that abnormal promoter hypermethylation profiling of cancer-associated genes may be a tumor-specific marker, and this alteration is often enhanced during disease progression, thus making it readily detectable in clinical specimens. In addition, hypermethylation constitutes a positively detectable signal with a very high degree of specificity, even in presence of a vast excess of unmethylated DNA. Detection of TSG hypermethylation has been detected in nipple aspirate [70] as well as in ductal lavage [71]. Jeronimo *et al.* [72] demonstrated that detection of gene hypermethylation in FNA washings by MSP is feasible. It is important that the target epigenetic alteration is cancer-specific and not present in normal or benign cells. Krassenstein *et al.* [70] using a hypermethylation panel of 6 genes (GSTP1, RARβ2, p16, p14, RASSF1A, and death-associated protein kinase) provided 100% coverage in nipple aspirate fluid samples. Jeronimo *et al.* did not select the cancer-specific targets, the methylated gene pattern could not discriminate between benign and malignant breast lesions in their study [72].

Serum or plasma is a more readily accessible bodily fluid, and provision of a specimen does not require the presence of a specialist. DNA is known to be released into serum/plasma, which is enriched from tumor DNA in cancer patients [73]. Importantly, tumor cell-specific DNA alterations in serum were not limited to patients with metastatic ancer but were also present in serum from patients with early or organ-confined tumors. Neoplastic DNA in the serum most likely arises from cells that have left the site of the primary lesion and have invaded the circulatory system but lack he capacity of metastasis to new organs or may be released from the primary tumor as free DNA from nonviable (apoptotic) neoplastic cells [1, 74]. Dulaimi *et al.* demonstrated tumor cell specific promoter hypermethylation of APC, RASSF1A, and DAP-kinase is present in ductal carcinoma *in situ*, lobular carcinoma *in situ*, and all grades and stages of invasive breast cancer. Hypermethylation can be detected by MSP analysis in serum DNA from patients with preinvasive and early-stage breast cancer amenable to cure.

Hypermethylation as a Marker for Histological Classification

According to the World Health Organization Classification, invasive breast carcinoma is subclassified as invasive ductal carcinoma, invasive lobular carcinoma and other special types. Ductal (over 80%) and lobular (about 10%) cancers are the two largest morphological subgroups of invasive breast cancer. Special subtypes include medullary carcinoma, mucinous carcinoma, papillary carcinoma, tubular carcinoma, adenoid cystic carcinoma, apocrine carcinoma, carcinoma with metaplasia, squamous type, spindle-cell type, cartilaginous and osseous type, mixed type, and others [75]. Because conventional histological behavior of specific cancers could not predict biological behavior, classifying cancers on a molecular level is urgently needed. Some researchers addressed the question whether different histological subtypes of cancer originating in the same organ differ in the epigenetic inactivation of essential genes.

Osada *et al.* [76] discovered a strong association between small lung cell cancer and 14-3-3δ hypermethylation, which was rarely found in non-small cell lung cancer. Graff *et al.* [77] found different methylation frequencies for the E-cadherin gene in different subtypes of thyroid carcinoma. Plachot *et al.* [78] investigated the effect of DNA methylation on the regulation of mammary acinar differention. They found that acinar morphogenesis was accompanied with chromatin remodeling, as shown by alterations in histone 4 acetylation, heterozhromatin 1 protein, and histone 3 methylated on lysine 9, and with an increase in expression of MeCP2. The results indicate that DNA methylation may act at least on two critical aspects of mammary epithelial differentiation, including tissue polarity and also potentially progenitor capabilities. Therefore, methylation pattern may be related with morphological phenotype in breast cancer.

Lehmann *et al.* [79] demonstrated that promoter hypermethylation of death-associated protein (DAP) kinase was found to be significantly higher in invasive lobular carcinoma compared to invasive ductal carcinoma. Fackler *et al.* [80] reported that Twist was hypermethylated less often in invasive lobular carcinoma than in invasive ductal carcinoma. Bae *et al.* [81] evaluated methylation status of 12 different genes in a series of 109 invasive breast tumors representing the ductal, lobular, and mucinous histologic subtypes using MSP. The selected genes include 10 genes that are commonly methylated in breast cancer (RASSF1A, HIN1, Twist, Cyclin D2, RARβ2, THRβ, E-cadherin, ER, BRCA1, and GSTP1), as well as two genes that are infrequently methylated in breast cancer (BAX and RB). As the results, lobular cancers and mucious cancers, which often have relatively low levels of chromosomal changes, have higher overall frequencies of hypermethylation than ductal cancers (49% in lobular and mucinous versus 40% in ductal), but there is a relatively unimodal distribution of methylation frequency of all three histologic types. Only BRCA1 shows a frequency of methylation that is significantly different among the various histologic categories, which is consistent an earlier report [53]. These findings suggest that methylation patterns of selected genes could contribute to the molecular classification of breast cancer. Differential methylation hybridization methods, including CpG island microarrays [66, 82, 83] are available to do epigenome-wide analysis. Analysis of more breast cancer cases and consideration of large amount of genes could help to define classes of breast cancer based on patterns of aberrant gene promoter methylation.

Hypermethylation as a Marker for Progression Evaluation

Another potential use of DNA methylation is in the classification of tumors depending on their methylation status. Such classification might be of use in determining patient prognosis. LATS1 and LATS2 are tumor suppressor genes implicated in the regulation of cell cycle. Hypermethylation of the promoter regions of LATS1 and LATS2 likely plays an important role in the down-regulation of their mRNA levels in breast cancers, and breast cancers with a decreased expression of LATS1 or LATS2 mRNA levels have a biologically aggressive phenotype [84]. Huang *et al.* developed an array-based methylation detection technology, called differential methylation hybridization (DMH), allowing for genome-wide screening of CpG island hypermethylation in breast cancer cell lines [82] and primary tumors [85]. Using DMH, they found that poorly differentiated tumors appeared to exhibit more hypermethylated CpG islands than their moderately or well-differentiated counterparts [85].

Systematic analysis of laser-microdissected ductal carcinoma *in situ* (DCIS) as well as epithelial hyperplasia and intraductal papillomas was performed by Lehamann *et al.* [62]. Four growth regulatory genes, p16, RASSF1, cyclin D2 and 14-3-3δ were investigated. They found that aberrant promoter methylation was already prevalent in pure intraductal carcinoma with different frequencies and different methylation levels for the four genes analyzed. RASSF1A and 14-3-3δ promoter methylation was also demonstrated in epithelial hyperplasia and intraductal papillomas. However, aberrant methylation of cyclinD2 and p16 was restricted to cancerous epithelium, and increased methylation of the cyclinD2 gene was significantly associated with a higher van Nuys grade.

The estrogen receptor (ER) gene and E-cadherin gene have been implicated frequently in the initiation and/or progression of human breast cancer. Loss of expression of either gene has been associated with poorly differentiated tumors and poorer prognosis [86, 87]. Loss of ER expression, as well as loss of E-cadherin expression, has been associated with aberrant 5'CpG methylation in breast cancers [88, 89]. Nass et al. [90] have surveyed 111 breast ductal carcinomas for aberrant methylation of ER and E-cadherin 5' CpG islands. Hypermethylation of either CpG island was evident prior to invasion in 30% of ductal carcinoma *in situ* lesions and increased significantly to nearly 60% in metastatic lesions. Coincident methylation of both CpG islands also increased significantly from 20% in ductal carcinoma *in situ* to nearly 50% in metastatic lesions. The increase in the coincident methylation of both CpG islands suggests that malignant progression of human breast cancer involves an accumulation of epigenetic events. The transcription factor activator protein-2α (AP-2α) is a tumor suppressor protein that can be lost during tumor progression. Recently, Douglas *et al.* [91] found that hypermethylation of a small CpG-rich region correlates with loss of AP-2α expression during progression of breast cancer. ER and E-cadherin are the down-stream factors of AP-2α, loss of AP-2α function might jield only a partial loss of tumor suppressor function for the ER and E-cadherin genes. Since ER and E-cadherin could also be attacked directly by local promoter change, coincident methylation of AP-2α, ER and E-cadherin may drive cancer invasion and progression.

The presence of methylation in certain genes within a tumour might be useful to identify more aggressive subtypes of cells. Such patterns of methylation in breast tumours may finally provide vital information for the individualisation of treatment. In addition, DNA methylation could be detected in bloodstream of breast cancer patients. Muller *et al.* [92] addressed whether DNA methylation in serum could be used as prognostic marker for breast cancer patients. Five genes (ESR1, APC, HSD17B4, HIC1, and RASSF1A) were selected to be candidate prognostic factors for breast cancer using a high-throughput MethyLight assay. Multivariate analysis showed methylated RASSF1A and/or APC serum DNA are independently associated with poor outcome.

Because neoadjuvant therapy has become increasingly popular as primary medical treatment of breast cancer, much attraction is focused on grading tumors on FNA cytology. Such grading would allow assessment of the tumor *in situ*, so the most-suitable treatment could be selected before primary surgery, and morbidity associated with overtreatment of low-grade tumors could be avoided. We developed a semiquatitative grading system for invasive ductal carcinoma on FNAC recently [93]. Without our method, tumor differentiation and progression could be evaluated preoperatively (Figure 5. Using FNA samples, combined analysis of cytological grading with certain genes methylation pattern may contribute to defined evaluation of tumor progression before operation.

Hypermethylation as a Predictor of Response to Treatment

The presence of estrogen receptor (ER) and/or progesterone receptor (PR) is an important diagnostic feature of breast cancer, and reflective of disease etiology. Over the three decades, the estrogen receptor antagonist tamoxifen has been mainstay of hormonal therapy, approximately 30% of patients treated with this agent still die within 10 years of follow-up treatment, and relapses can occur for more than 20 years following therapy [94]. The cause of treatment failure in many breast cancer patients receiving tamoxifen is resistance to it. Hormone receptor (HR) status, defined as ER and/or PR positivity, has been shown to predict response to tamoxifen treatment [95]. ERs comprose two subtypes, ER α and ER β. ER β was often coexpressed with ER α and sometimes inversely correlated PR. Recent studies have shown that hypermethylation appears to play a role in inactivation of these genes. Southern blot and MSP analyses revealed that ER α was unmethylated in normal breast tissue and ER-positive tumor cell lines, such as MCF-7, T47-D and ZR75-1; it is methylated in about 50%

of unselected primary breast cancers and most ER-negative cancer cell lines, *e.g.* MDA-MB-231, MDA-MB-435, MDA-MB-468, Hs578t and MCF-7/Adr. The ER α island methylation is associated with reduced or absent ER mRNA expression. PR gene methylation has been demonstrated by Southern analysis in about 40% of PR-negative breast tumors and several PR-negative breast cancer cell lines such as MDA-MB-231, MDA-MB-468, MDA-MB-435 and Hs578t. Chang *et al.* [96] investigated the methylation patterns of the ER genes in the tamoxifen-resistant tumors. They used bisulfite genomic sequencing and reverse transcriptase PCR to determine the methylation patterns and mRNA expression of the two ER genes, α and β, from control and tamoxifen-resistant tissues chosen by an age-matched sampling method. The methylation rate of the ER β but not the ER α in the control group was significantly higher than in its counterpart. Among the methylated tumors mean methylation density of ER α and ER β in the resistant cases was significantly elevated. The tamoxifen-resistant tumors showed more dense methylation of the ER β gene than control tumors.

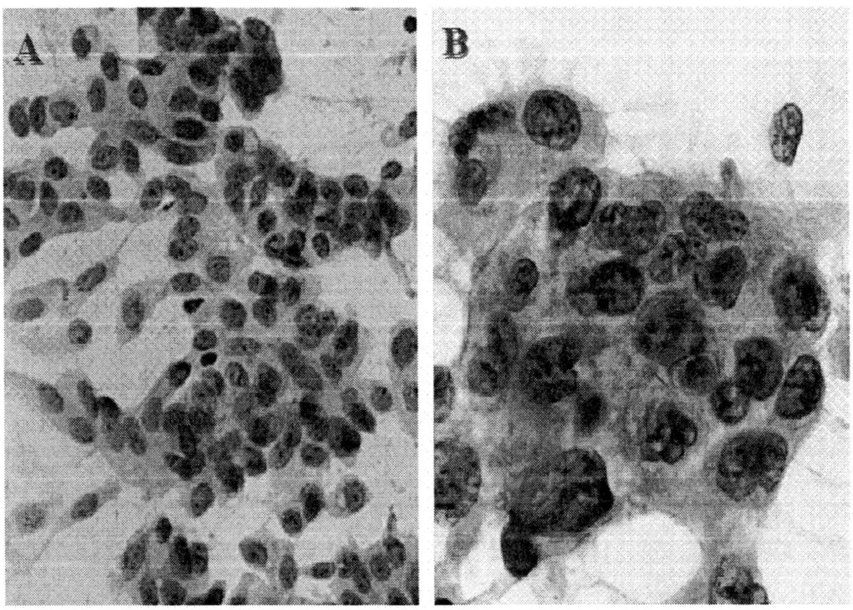

Figure 5. Cytological smear of invasive ductal carcinoma (Papanicolaou stain x40). (A) a well differentiated ductal carcinoma. (B) a poorly differentiated ductal carcinoma.

Rectinoic acid receptor β is also a member of the nuclear receptor superfamily. Loss of RAR β2 has been observed in solid tumor cells, including lung carcinoma, squamous cell carcinoma, and breast cancer. These findings suggest that the specific loss of RAR β2 expression may be an important event during tumorigenesis. This hypothesis is supported by the observation that introduction of RAR β into retinoic acid-insensitive breast cancer cell lines restored retinoic acid responsiveness to growth inhibition and induction of apoptosis. To further examine the hypothesis that the chemopreventive effects of retinoids are due to RAR β2 induction, Sun *et al.* [97] blocked ATRA-induced RAR β expression in H157 cells using a retroviral vector harboring multiple copies of antisense RAR β2 sequences. Antisense RAR β-transfected cells shows not only decreased expression of ATRA-induced RAR β protein but also reduced ATRA-induced RARE binding activity and transactivation. Importantly, all antisense RAR β transfectants of H157 cells were less responsive than vector-retinoids ATRA and Ch55 *in vitro*. These results demonstrated that RAR β induction might play an important role in mediating growth inhibitory effects of retinoids in cancer cells. DNA methylation of the RAR β2 has been demonstrated in human cancer cell lines and tumors [98, 99], methylation assay of RAR β2 may provide critical information for retinoid response in breast cancer.

Combined use of epigenetic drugs with agents that target reactivated genes—in effect using epigenetic drugs as cellular sensitizers is a promising strategy for cancer treatment. Restoration of response to retinoid signaling using this approach has been observed in acute myeloid leukemia and in colon and breast cancers [100-102]. This approach could be used to reactivate other genes silenced in precancerous tissues that are targets for established chemopreventive agents such as ER gene described above. Recently, Fiegl et al. [103] analyzed *RASSF1A* DNA methylation in pretherapeutic sera and serum samples collected 1 year after surgery from 148 patients with breast cancer who were receiving adjuvant tamoxifen; 19.6% and 22.3% of patients with breast cancer showed *RASSF1A* DNA methylation in their pretherapeutic and 1-year-after serum samples, respectively. *RASSF1A* methylation 1 year after primary surgery (and during adjuvant tamoxifen therapy) was an independent predictor of poor outcome. Based on their findings, measurement of serum DNA methylation allows adjuvant systemic treatment to be monitored for efficacy: disappearance of *RASSF1A* DNA methylation in serum throughout treatment with tamoxifen indicates a response, whereas persistence or new appearance means resistance to adjuvant tamoxifen treatment. Therefore, serum DNA methylation such as *RASSF1A* is an easy means of detecting patients undergoing adjuvant treatment who need advanced treatment.

Epigenetic Modulators for Breast Cancer Treatment

Unlike genetic alterations such as mutations, epigenetic changes such as DNA methylation are potentially reversible. Methylation causes the inactivation of numerous genes that are important in the development of most tumor types including breast cancer, therefore, epigenetic changes are attractive target for cancer therapy. 5-Azacytidine and 5-aza-2'-deoxycytidine (5-Aza-CdR) (Figure 6), nucleoside analog inhibitors of DNMTs, have been widely used to reverse abnormal DNA hypermethylation and restore silenced gene expression. In one of our previous studies [104], the effect of 5-Aza-CdR RAR β2 restoration was investigated in the MRK-nu-1 human female breast cancer cell line. Changes of the RAR β2 methylation status were assessed by MSP. Reverse transcription PCR was used to evaluate RAR β2 restoration. Cell cycling and growth inhibition were studied using flow cytometric analysis of DNA content and CellTiter 96 AQueous non-radioactive cell proliferation assay, respectively. 5-Aza-CdR treatment resulted in complete demethylation of the RAR β2 gene. RAR β2 restoration was accompanied by cell cycle arrest (increase in the G0/G1 and decrease in the S- and G2/M-phases) and time-dependent growth inhibition. Nucleoside analog inhibitors of DNMTs have many potential side effects, such as myelotoxicity, mutagenesis, and tumorigenesis. A few non-nucleoside inhibitors of DNA methyltransferases have been reported, including the anti-arrhythmia drug, procainamide. Following this need to find new demethylating agents, Villar-Garea *et al.* [105] have tested the potential use of procaine, an anesthetic drug related to procainamide. Using the MCF-7 breast cancer cell line, they found that procaine is a DNA-demethylating agent that produces a 40% reduction in 5-methylcytosine DNA content as determined by high-performance capillary electrophoresis or total DNA enzyme digestion. Procaine can also demethylate densely hypermethylated CpG islands, such as those located in the promoter region of the RAR β2 gene, restoring gene expression of epigenetically silenced genes. Non-nucleoside analog DNA methyltransferase inhibitors such as procainamide are attractive alternative for possible clinical use. Very recently, Grunt et al.[106] reported that other reagents may also have demethylation activity. They demonstrated that the EGFR inhibitor PD153035 not only blocked activation of EGFR and inhibited cell growth, but also stimulated RAR-beta expression in MDA-MB-468 breast and OVCAR-3 ovarian carcinoma cells. Upregulation of RAR-beta by PD153035 was confirmed by real-time reverse transcription-polymerase chain reaction. In contrast, expression of other retinoid receptors and of estrogen receptor-alpha was not affected. PD153035-mediated re-induction of RAR-beta was associated with demethylation of the RAR-beta2 gene promoter P2 as demonstrated by methylation-specific polymerase chain reaction. These novel results on the ErbB/retinoid receptor cross-talk may be useful for designing future anticancer combination regimens.

As described above, histone deacetylase activity is important in the transcriptional repression of methylated sequences. Histone deacetylase inhibitors induce cell-cycle arrest, apoptosis, and/or differentiation in transformed cells *in vitro* and suppress the growth of a wide variety of solid tumor xenografts with minimal toxicity. Several structural classes of histone deacetylase inhibitors have been identified, including short-chain fatty acids (butyric and valproic acids), hydroxamic acids (suberoylanilide hydroxamic acid, *m*-carboxycinnamic acid bishydroxamide, trichostatin A, oxamflatin, and pyroxamide), tetrapeptides (depsipeptide), and benzamides (MS-275, which stands for *N*-(2-aminophenyl)-4-[*N*-(pyridin-3-yl-methoxycarbonyl)aminomethyl]benzamide and CI-994, which stands for *N*-acetyldinaline). Keen *et al.* [107] demonstrated that combining treatment with 5-aza-2'-deoxycytidine and the histone deacetylase (HDAC) inhibitor resulted in expression of functional ER mRNA and protein. Suberoylanilide hydroxamic acid decreased the incidence and multiplicity of *N*-methyl-*N*-nitrosourea-induced rat mammary tumors; unlike 5-aza-2'-deoxycytidine, inhibitory effects were manifest at stages after initiation. Suberoylanilide hydroxamic acid also inhibited hyperplastic nodule formation in response to 7,12-dimethylbenz[*a*]anthracene treatment in mouse mammary gland organ culture and the multiplicity of carcinogen-induced lung tumors in mice [108]. Very recently, Sharma *et al.* [109] demonstrated that inhibitors of DNA DNMT and histone deacetylases (HDAC) synergistically activate the methylated ER gene promoter in ER negative MDA-MB-231 human breast cancer cells. Chromatin immunoprecipitation (ChIP) was used to examine the chromatin status and repressor complex associated with silenced ER and changes in the key regulatory factors during reactivation by inhibitors of DNMT (5-aza-2'deoxycytidine or 5-aza-dC) and HDAC (trichostatin A or TSA). The silencing of ER due to CpG hypermethylation correlates with binding of specific methyl-binding proteins, DNA methyltransferases and HDAC proteins. Inhibition of HDAC activity by TSA results in the accumulation of hyperacetylated core histones. The activation of ER gene expression by 5aza-dC also involves the release of the repressor complex involving various methyl binding proteins, DNMTs, and HDAC1. HDAC and DNMT inhibitors modulate histone methylation at H3-K9 and H3-K4 to form a more open chromatin structure necessary for reactivation of silenced ER transcription. Combining demethylating agents and histone deacetylase inhibitors may provide a means to reduce the side effects associated with the former drugs. Synergy between 5-aza-2'-deoxycytidine and trichostatin A has been achieved in reactivating methylation-silenced tumor suppressor genes [110].

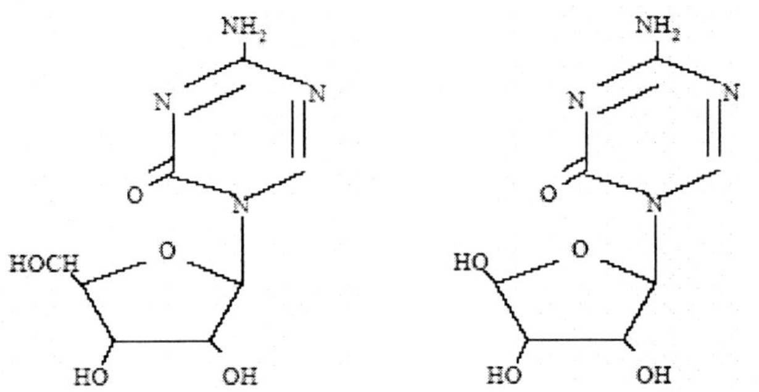

5-Azacytidine 5-Aza-2'-deoxycytidine

Figure 6. Structure of 5-Azacytidine and 5-Aza-2'-deoxycytidine

Conclusion

DNA methylation is an epigenetic modification with an important role in the control of gene expression in human cancer. Within the past few years, the number of known DNMT enzymes, the number of genes known to be targeted by methylation have increased significantly and the mechanisms by which DNA methylation leads to transcriptional inactivation have been greatly elucidated. Aberrant promoter methylation of cancer-related genes are not randomly distributed in breast cancers [111], therefore, breast cancer could be grouped based on methylation pattern. DNMTs such as DNMT1, 3A, 3B are differentially distributed in breast cancer [112], the relative roles of the individual DNMTs in individual gene expression remain unclear. The reason that hypomethylation and hypermethylation co-exist in breast cancer tumors need further research. Breast cancer cells typically contain somatically altered nucleic acid, characterized by mutation, amplification, or deletion of critical genes. In addition, the nucleic acid from breast cancer cells often displays epigenetic changes, e.g. DNA methylation. Currently, there is a lack of both evidence and consensus with regard to the timing of genetic versus epigenetic perturbations in tumorigenesis and plausible non-mutually exclusive mechanisms can be proposed in either case. The timing of the methylation and gene inactivation process with respect to tumor formation remains to be properly defined.

DNA methylation detection can serve as a promising tumor biomarker of early diagnosis. Diagnostic approaches based on identification of epigenetic changes in specific genes are likely to allow implementation of early detection strategies and novel therapeutic approaches. Targeting these early changes might lead to more effective cancer treatment. However, very few markers have been demonstrated with high sensitivity and specificity. Identifying new aberrantly methylated genes using high-throughput microarray technology should be performed toward tumor marker selection. Increased interests are focused on using novel approaches to reversing tumor-specific methylation.

Combining demethylating agents and histone deacetylase inhibitors may provide a means to reduce the side effects associated with the former drugs. Synergy between 5-aza-2'-deoxycytidine and trichostatin A has been achieved in reactivating methylation-silenced tumor suppressor genes[110]. Other promising strategies include the use of the combined use of epigenetic drugs with agents that target reactivated genes, in this way, epigenetic drugs are used as cellular sensitizers. Restoration of response to retinoid signaling using this approach has been observed in many tumor types including breast cancer [102, 113, 114]. In addition, a combination of DNMT inhibitors and histone deacetylation inhibitors may be an alterative attractive strategy for the treatment of breast cancer patients.

References

[1] Sidransky, D., Emerging molecular markers of cancer. *Nat Rev Cancer*, 2002. *2*(3): p. 210-9.

[2] Laird, P.W., The power and the promise of DNA methylation markers. *Nat Rev Cancer*, 2003. *3*(4): p. 253-66.

[3] Ballestar, E. and M. Esteller, The impact of chromatin in human cancer: linking DNA methylation to gene silencing. *Carcinogenesis*, 2002. *23*(7): p. 1103-9.

[4] Takai, D. and P.A. Jones, Comprehensive analysis of CpG islands in human chromosomes 21 and 22. *Proc Natl Acad Sci U S A*, 2002. *99*(6): p. 3740-5.

[5] Robertson, K.D., *DNA* methylation and chromatin - unraveling the tangled web. *Oncogene*, 2002. *21*(35): p. 5361-79.

[6] Costello, J.F. and C. Plass, Methylation matters. *J Med Genet*, 2001. *38*(5): p. 285-303.

[7] Yoder, J.A. and T.H. Bestor, A candidate mammalian DNA methyltransferase related to pmt1p of fission yeast. *Hum Mol Gene*t, 1998. *7*(2): p. 279-84.

[8] Bestor, T., et al., Cloning and sequencing of a cDNA encoding DNA methyltransferase of mouse cells. The carboxyl-terminal domain of the mammalian enzymes is related to bacterial restriction methyltransferases. *J Mol Bio*l, 1988. *203*(4): p. 971-83.

[9] Agoston, A.T., et al., Increased protein stability causes DNA methyltransferase 1 dysregulation in breast cancer. *J Biol Chem*, 2005.

[10] Okano, M., S. Xie, and E. Li, Cloning and characterization of a family of novel mammalian DNA (cytosine-5) methyltransferases. *Nat Genet*, 1998. *19*(3): p. 219-20.

[11] Robertson, K.D., et al., The human DNA methyltransferases (DNMTs) 1, 3a and 3b: coordinate mRNA expression in normal tissues and overexpression in tumors. *Nucleic Acids Res*, 1999. *27*(11): p. 2291-8.

[12] Xie, S., et al., Cloning, expression and chromosome locations of the human DNMT3 gene family. *Gene*, 1999. *236*(1): p. 87-95.

[13] Okano, M., et al., DNA methyltransferases Dnmt3a and Dnmt3b are essential for de novo methylation and mammalian development. *Cell*, 1999. *99*(3): p. 247-57.

[14] Xu, G.L., et al., Chromosome instability and immunodeficiency syndrome caused by mutations in a DNA methyltransferase gene. *Nature*, 1999. *402*(6758): p. 187-91.

[15] Ramsahoye, B.H., et al., Non-CpG methylation is prevalent in embryonic stem cells and may be mediated by DNA methyltransferase 3a. *Proc Natl Acad Sci U S A*, 2000. *97*(10): p. 5237-42.

[16] Mizuno, S., et al., Expression of DNA methyltransferases DNMT1, 3A, and 3B in normal hematopoiesis and in acute and chronic myelogenous leukemia. *Blood*, 2001. *97*(5): p. 1172-9.

[17] Ordway, J.M. and T. Curran, Methylation matters: modeling a manageable genome. *Cell Growth Differ*, 2002. *13*(4): p. 149-62.

[18] Howell, C.Y., et al., Genomic imprinting disrupted by a maternal effect mutation in the Dnmt1 gene. *Cell,* 2001. *104*(6): p. 829-38.

[19] Mohandas, T., R.S. Sparkes, and L.J. Shapiro, Reactivation of an inactive human X chromosome: evidence for X inactivation by DNA methylation. *Science*, 1981. *211*(4480): p. 393-6.

[20] Nan, X., et al., Transcriptional repression by the methyl-CpG-binding protein MeCP2 involves a histone deacetylase complex. *Nature*, 1998. *393*(6683): p. 386-9.

[21] Lewis, J.D., et al., Purification, sequence, and cellular localization of a novel chromosomal protein that binds to methylated DNA. *Cell,* 1992. *69*(6): p. 905-14.

[22] Hendrich, B. and A. Bird, Identification and characterization of a family of mammalian methyl-CpG binding proteins. *Mol Cell Biol*, 1998. *18*(11): p. 6538-47.

[23] Magdinier, F., et al., Regional methylation of the 5' end CpG island of BRCA1 is associated with reduced gene expression in human somatic cells. *Faseb J*, 2000. *14*(11): p. 1585-94.

[24] Ballestar, E., et al., Methyl-CpG binding proteins identify novel sites of epigenetic inactivation in human cancer. *Embo J*, 2003. *22*(23): p. 6335-45.

[25] Jones, P.L., et al., Methylated DNA and MeCP2 recruit histone deacetylase to repress transcription. *Nat Genet*, 1998. *19*(2): p. 187-91.

[26] Turner, B.M., Cellular memory and the histone code. *Cell,* 2002. *111*(3): p. 285-91.

[27] Fuks, F., et al., The methyl-CpG-binding protein MeCP2 links DNA methylation to histone methylation. *J Biol Chem*, 2003. *278*(6): p. 4035-40.

[28] Robertson, K.D., et al., DNMT1 forms a complex with Rb, E2F1 and HDAC1 and represses transcription from E2F-responsive promoters. *Nat Genet*, 2000. *25*(3): p. 338-42.

[29] Burgers, W.A., F. Fuks, and T. Kouzarides, DNA methyltransferases get connected to chromatin. *Trends Genet*, 2002. *18*(6): p. 275-7.

[30] Matzke, M., A.J. Matzke, and J.M. Kooter, RNA: guiding gene silencing. *Science*, 2001. *293*(5532): p. 1080-3.
[31] Kawasaki, H. and K. Taira, Induction of DNA methylation and gene silencing by short interfering RNAs in human cells. *Nature*, 2004. *431*(7005): p. 211-7.
[32] Baylin, S.B., et al., Aberrant patterns of DNA methylation, chromatin formation and gene expression in cancer. *Hum Mol Genet*, 2001. *10*(7): p. 687-92.
[33] Feinberg, A.P., et al., Reduced genomic 5-methylcytosine content in human colonic neoplasia. *Cancer Res*, 1988. *48*(5): p. 1159-61.
[34] Narayan, A., et al., Hypomethylation of pericentromeric DNA in breast adenocarcinomas. *Int J Cancer*, 1998. *77*(6): p. 833-8.
[35] Nass, S.J., et al., Expression of DNA methyl-transferase (DMT) and the cell cycle in human breast cancer cells. *Oncogene*, 1999. *18*(52): p. 7453-61.
[36] De Marzo, A.M., et al., Abnormal regulation of DNA methyltransferase expression during colorectal carcinogenesis. *Cancer Res*, 1999. *59*(16): p. 3855-60.
[37] Szyf, M., Targeting DNA methylation in cancer. *Ageing Res Rev*, 2003. *2*(3): p. 299-328.
[38] Magewu, A.N. and P.A. Jones, Ubiquitous and tenacious methylation of the CpG site in codon 248 of the p53 gene may explain its frequent appearance as a mutational hot spot in human cancer. *Mol Cell Biol*, 1994. *14*(6): p. 4225-32.
[39] Ghazi, H., et al., Changes in the allelic methylation patterns of c-H-ras-1, insulin and retinoblastoma genes in human development. *Dev Suppl*, 1990: p. 115-23.
[40] Strand, M., et al., Destabilization of tracts of simple repetitive DNA in yeast by mutations affecting DNA mismatch repair. *Nature*, 1993. *365*(6443): p. 274-6.
[41] Prolla, T.A., et al., Tumour susceptibility and spontaneous mutation in mice deficient in Mlh1, Pms1 and Pms2 DNA mismatch repair. *Nat Genet*, 1998. *18*(3): p. 276-9.
[42] Lengauer, C., K.W. Kinzler, and B. Vogelstein, Genetic instability in colorectal cancers. *Nature*, 1997. *386*(6625): p. 623-7.
[43] Lengauer, C., K.W. Kinzler, and B. Vogelstein, Genetic instabilities in human cancers. *Nature*, 1998. *396*(6712): p. 643-9.
[44] Ferguson, A.T., et al., High frequency of hypermethylation at the 14-3-3 sigma locus leads to gene silencing in breast cancer. *Proc Natl Acad Sci U S A*, 2000. *97*(11): p. 6049-54.
[45] Almeida, A., et al., Hypomethylation of classical satellite DNA and chromosome instability in lymphoblastoid cell lines. *Hum Genet*, 1993. *91*(6): p. 538-46.
[46] Chen, R.Z., et al., DNA hypomethylation leads to elevated mutation rates. *Nature*, 1998. *395*(6697): p. 89-93.
[47] Ji, W., et al., DNA demethylation and pericentromeric rearrangements of chromosome *1. Mutat Res*, 1997. *379*(1): p. 33-41.
[48] Kane, M.F., et al., *Methylation of the hMLH1 promoter correlates with lack of expression of hMLH1 in sporadic colon tumors and mismatch repair-defective human tumor cell lines.*, 1997. *57*(5): p. 808-11.
[49] Esteller, M., et al., hMLH1 promoter hypermethylation is an early event in human endometrial tumorigenesis. *Am J Pathol*, 1999. *155*(5): p. 1767-72.

[50] Cavalieri, E.L., et al., Molecular origin of cancer: catechol estrogen-3,4-quinones as endogenous tumor initiators. *Proc Natl Acad Sci U S A*, 1997. *94*(20): p. 10937-42.

[51] Qu, G.Z., et al., Frequent hypomethylation in Wilms tumors of pericentromeric DNA in chromosomes 1 and 16. *Cancer Genet Cytogenet*, 1999. *109*(1): p. 34-9.

[52] Frommer, M., et al., A genomic sequencing protocol that yields a positive display of 5-methylcytosine residues in individual DNA strands. *Proc Natl Acad Sci U S A*, 1992. *89*(5): p. 1827-31.

[53] Herman, J.G., et al., Methylation-specific PCR: a novel PCR assay for methylation status of CpG islands. *Proc Natl Acad Sci U S A*, 1996. *93*(18): p. 9821-6.

[54] Li, L.C. and R. Dahiya, MethPrimer: designing primers for methylation PCRs. Bioinformatics, 2002. *18*(11): p. 1427-31.

[55] Rozen, S. and H. Skaletsky, Primer3 on the WWW for general users and for biologist programmers. *Methods Mol Biol*, 2000. *132*: p. 365-86.

[56] Palmisano, W.A., et al., Predicting lung cancer by detecting aberrant promoter methylation in sputum. *Cancer Res*, 2000. *60*(21): p. 5954-8.

[57] Fraga, M.F. and M. Esteller, *DNA methylation: a profile of methods and applications. Biotechniques*, 2002. *33*(3): p. 632, 634, 636-49.

[58] Harrison, J., C. Stirzaker, and S.J. Clark, Cytosines adjacent to methylated CpG sites can be partially resistant to conversion in genomic bisulfite sequencing leading to methylation artifacts. *Anal Biochem,* 1998. *264*(1): p. 129-32.

[59] Rein, T., et al., Absence of an unusual "densely methylated island" at the hamster dhfr ori-beta. *J Biol Chem*, 1997. *272*(15): p. 10021-9.

[60] Eads, C.A., et al., MethyLight: a high-throughput assay to measure DNA methylation. *Nucleic Acids Res*, 2000. *28*(8): p. E32.

[61] Fackler, M.J., et al., Quantitative multiplex methylation-specific PCR assay for the detection of promoter hypermethylation in multiple genes in breast cancer. *Cancer Res*, 2004. *64*(13): p. 4442-52.

[62] Lehmann, U., et al., Quantitative assessment of promoter hypermethylation during breast cancer development. *Am J Pathol*, 2002. *160*(2): p. 605-12.

[63] Cottrell, S.E. and P.W. Laird, Sensitive detection of DNA methylation. *Ann N Y Acad Sci*, 2003. *983*: p. 120-30.

[64] Tompa, R., et al., Genome-wide profiling of DNA methylation reveals transposon targets of CHROMOMETHYLASE3. *Curr Biol*, 2002. *12*(1): p. 65-8.

[65] Gitan, R.S., et al., Methylation-specific oligonucleotide microarray: a new potential for high-throughput methylation analysis. *Genome Res*, 2002. *12*(1): p. 158-64.

[66] Yan, P.S., et al., Dissecting complex epigenetic alterations in breast cancer using CpG island microarrays. *Cancer Res*, 2001. *61*(23): p. 8375-80.

[67] Hatada, I., et al., A microarray-based method for detecting methylated loci. *J Hum Genet*, 2002. *47*(8): p. 448-51.

[68] Yang, Q., et al., Two-hit inactivation of FHIT by loss of heterozygosity and hypermethylation in breast cancer. *Clin Cancer Res*, 2002. *8*(9): p. 2890-3.

[69] Lewis, C.M., et al., Promoter hypermethylation in benign breast epithelium in relation to predicted breast cancer risk. *Clin Cancer Res*, 2005. *11*(1): p. 166-72.

[70] Krassenstein, R., et al., Detection of breast cancer in nipple aspirate fluid by CpG island hypermethylation. *Clin Cancer Res*, 2004. *10*(1 Pt 1): p. 28-32.

[71] Evron, E., et al., Detection of breast cancer cells in ductal lavage fluid by methylation-specific PCR. *Lancet*, 2001. *357*(9265): p. 1335-6.

[72] Jeronimo, C., et al., Detection of gene promoter hypermethylation in fine needle washings from breast lesions. *Clin Cancer Res*, 2003. *9*(9): p. 3413-7.

[73] Leon, S.A., et al., Free DNA in the serum of cancer patients and the effect of therapy. *Cancer Res*, 1977. *37*(3): p. 646-50.

[74] Dulaimi, E., et al., Tumor suppressor gene promoter hypermethylation in serum of breast cancer patients. *Clin Cancer Res*, 2004. *10*(18 Pt 1): p. 6189-93.

[75] Berg, J.W. and R.V. Hutter, Breast cancer. *Cancer*, 1995. *75*(1 Suppl): p. 257-69.

[76] Osada, H., et al., Frequent and histological type-specific inactivation of 14-3-3sigma in human lung cancers. *Oncogene*, 2002. *21*(15): p. 2418-24.

[77] Graff, J.R., et al., Distinct patterns of E-cadherin CpG island methylation in papillary, follicular, Hurthle's cell, and poorly differentiated human thyroid carcinoma. *Cancer Res*, 1998. *58*(10): p. 2063-6.

[78] Plachot, C. and S.A. Lelievre, DNA methylation control of tissue polarity and cellular differentiation in the mammary epithelium. *Exp Cell Res*, 2004. *298*(1): p. 122-32.

[79] Lehmann, U., et al., Promoter hypermethylation of the death-associated protein kinase gene in breast cancer is associated with the invasive lobular subtype. *Cancer Res*, 2002. *62*(22): p. 6634-8.

[80] Fackler, M.J., et al., DNA methylation of RASSF1A, HIN-1, RAR-beta, Cyclin D2 and Twist in in situ and invasive lobular breast carcinoma. *Int J Cancer*, 2003. *107*(6): p. 970-5.

[81] Bae, Y.K., et al., Hypermethylation in histologically distinct classes of breast cancer. *Clin Cancer Res*, 2004. *10*(18 Pt 1): p. 5998-6005.

[82] Huang, T.H., M.R. Perry, and D.E. Laux, Methylation profiling of CpG islands in human breast cancer cells. *Hum Mol Genet*, 1999. *8*(3): p. 459-70.

[83] Shi, H., et al., Expressed CpG island sequence tag microarray for dual screening of DNA hypermethylation and gene silencing in cancer cells. *Cancer Res*, 2002. *62*(11): p. 3214-20.

[84] Takahashi, Y., et al., Down-regulation of LATS1 and LATS2 mRNA expression by promoter hypermethylation and its association with biologically aggressive phenotype in human breast cancers. *Clin Cancer Res*, 2005. *11*(4): p. 1380-5.

[85] Huang, S., et al., Characterization of the human and mouse WRN 3'-->5' exonuclease. *Nucleic Acids Res*, 2000. *28*(12): p. 2396-405.

[86] McGuire, W.L., Hormone receptors: their role in predicting prognosis and response to endocrine therapy. *Semin Oncol*, 1978. *5*(4): p. 428-33.

[87] Siitonen, S.M., et al., Reduced E-cadherin expression is associated with invasiveness and unfavorable prognosis in breast cancer. *Am J Clin Pathol*, 1996. *105*(4): p. 394-402.

[88] Lapidus, R.G., et al., Methylation of estrogen and progesterone receptor gene 5' CpG islands correlates with lack of estrogen and progesterone receptor gene expression in breast tumors. *Clin Cancer Res*, 1996. *2*(5): p. 805-10.

[89] Graff, J.R., et al., E-cadherin expression is silenced by DNA hypermethylation in human breast and prostate carcinomas. *Cancer Res*, 1995. *55*(22): p. 5195-9.

[90] Nass, S.J., et al., Aberrant methylation of the estrogen receptor and E-cadherin 5' CpG islands increases with malignant progression in human breast cancer. *Cancer Res*, 2000. *60*(16): p. 4346-8.

[91] Douglas, D.B., et al., Hypermethylation of a small CpGuanine-rich region correlates with loss of activator protein-2alpha expression during progression of breast cancer. *Cancer Res*, 2004. *64*(5): p. 1611-20.

[92] Muller, H.M., et al., DNA methylation in serum of breast cancer patients: an independent prognostic marker. *Cancer Res*, 2003. *63*(22): p. 7641-5.

[93] Taniguchi, E., et al., Cytologic grading of invasive breast carcinoma. Correlation with clinicopathologic variables and predictive value of nodal metastasis. *Acta Cytol*, 2000. *44*(4): p. 587-91.

[94] Eiermann, W., et al., Preoperative treatment of postmenopausal breast cancer patients with letrozole: A randomized double-blind multicenter study. *Ann Oncol*, 2001. *12*(11): p. 1527-32.

[95] Bardou, V.J., et al., Progesterone receptor status significantly improves outcome prediction over estrogen receptor status alone for adjuvant endocrine therapy in two large breast cancer databases. *J Clin Oncol*, 2003. *21*(10): p. 1973-9.

[96] Chang, H.G., et al., Tamoxifen-resistant breast cancers show less frequent methylation of the estrogen receptor beta but not the estrogen receptor alpha gene. *J Mol Med*, 2005. *83*(2): p. 132-9.

[97] Sun, S.Y., et al., Evidence that retinoic acid receptor beta induction by retinoids is important for tumor cell growth inhibition. *J Biol Chem*, 2000. *275*(22): p. 17149-53.

[98] Yang, Q., T. Sakurai, and K. Kakudo, Retinoid, retinoic acid receptor beta and breast cancer. *Breast Cancer Res Treat*, 2002. *76*(2): p. 167-73.

[99] Yang, Q., et al., Biallelic inactivation of retinoic acid receptor beta2 gene by epigenetic change in breast cancer. *Am J Pathol*, 2001. *158*(1): p. 299-303.

[100] Ferrara, F.F., et al., Histone deacetylase-targeted treatment restores retinoic acid signaling and differentiation in acute myeloid leukemia. *Cancer Res*, 2001. *61*(1): p. 2-7.

[101] Cote, S. and R.L. Momparler, Activation of the retinoic acid receptor beta gene by 5-aza-2'-deoxycytidine in human DLD-1 colon carcinoma cells. *Anticancer Drugs*, 1997. *8*(1): p. 56-61.

[102] Widschwendter, M., et al., Methylation and silencing of the retinoic acid receptor-beta2 gene in breast cancer. *J Natl Cancer Inst*, 2000. *92*(10): p. 826-32.

[103] Fiegl, H., et al., Circulating tumor-specific DNA: a marker for monitoring efficacy of adjuvant therapy in cancer patients. *Cancer Res*, 2005. *65*(4): p. 1141-5.

[104] Yang, Q., et al., 5-aza-2'-deoxycytidine induces retinoic acid receptor beta 2 demethylation, cell cycle arrest and growth inhibition in breast carcinoma cells. *Anticancer Res*, 2002. *22*(5): p. 2753-6.

[105] Villar-Garcia, A., et al., Procaine is a DNA-demethylating agent with growth-inhibitory effects in human cancer cells. *Cancer Res*, 2003. *63*(16): p. 4984-9.

[106] Grunt, T.W., et al., An EGF receptor inhibitor induces RAR-beta expression in breast and ovarian cancer cells. *Biochem Biophys Res Commun*, 2005. *329*(4): p. 1253-9.

[107] Keen, J.C., et al., A novel histone deacetylase inhibitor, scriptaid, enhances expression of functional estrogen receptor alpha (ER) in ER negative human breast cancer cells in combination with 5-aza 2'-deoxycytidine. *Breast Cancer Res Treat*, 2003. *81*(3): p. 177-86.

[108] Desai, D., et al., Chemopreventive efficacy of suberoylanilide hydroxamic acid (SAHA) against 4-(methylnitrosamino)-1-(3-pyridyl)-1-butanone (NNK)-induced lung tumorigenesis in female A/J mice. *Anticancer Res*, 2003. *23*(1A): p. 499-503.

[109] Sharma, D., et al., Release of Methyl CpG Binding Proteins and Histone Deacetylase 1 from the Estrogen Receptor Alpha (ER) Promoter upon Reactivation in ER-negative Human Breast Cancer Cells. *Mol Endocrinol,* 2005.

[110] Cameron, E.E., et al., Synergy of demethylation and histone deacetylase inhibition in the re-expression of genes silenced in cancer. *Nat Genet*, 1999. *21*(1): p. 103-7.

[111] Parrella, P., et al., Nonrandom distribution of aberrant promoter methylation of cancer-related genes in sporadic breast tumors. *Clin Cancer Res*, 2004. *10*(16): p. 5349-54.

[112] Girault, I., et al., Expression analysis of DNA methyltransferases 1, 3A, and 3B in sporadic breast carcinomas. *Clin Cancer Res*, 2003. *9*(12): p. 4415-22.

[113] Sirchia, S.M., et al., Evidence of epigenetic changes affecting the chromatin state of the retinoic acid receptor beta2 promoter in breast cancer cells. *Oncogene*, 2000. *19*(12): p. 1556-63.

[114] Sirchia, S.M., et al., Endogenous reactivation of the RARbeta2 tumor suppressor gene epigenetically silenced in breast cancer. *Cancer Res*, 2002. *62*(9): p. 2455-61.

Index

A

B

C

D

E

F

I

K

L

M

N

O

P

Q

R

S

T